RHIC:
Optics measure, correct.

Xiaozhe Shen

To my parents.

Acknowledgments

The work of this dissertation would not have been accomplished if it were not for the help and support from many people. It is my pleasure to have an opportunity to express my gratitude at this page.

First of all, I would like to thank my thesis adviser, Prof. S. Y. Lee, for his earnest guidance, generous support and continued encouragement. I have benefited a lot from his profound knowledge and assiduous attitude towards research. It is him who introduces me to this interesting Ph.D. topic and provides me with numerous useful discussions. I am honored to be one of his students and would like to show him my deepest respect.

I would also like to thank many people for their help during my visit to the Collider-Accelerator Department (CAD) of Brookhaven National Lab (BNL) for my PhD study. Firstly, I would like to thank Dr. Wolfram Fischer and Dr. Mei Bai for their support and generosity. I would especially like to thank Mei for the continuous help and stimulating discussions, whether research related or not. I would also like to thank Dr. Simon White for many useful discussions and comments on my work and Dr. Steven Tepikian for his help on RHIC design model. I am grateful to Dr. Lingyun Yang from National Synchrotron Light SourceïjŇ Dr. Yun Luo, and Dr. Yue Hao for their help in computer simulations as well as invaluable discussions in general accelerator physics. I would also like to thank Dr. François Méot, Dr. Jorg Kewisch, Dr. Michiko Minty, Dr. Chuyu Liu, Dr. Yichao Jing, Dr. Stephen Peggs, Peter Oddo, Arthur Fernando, engineers in the beam instrumentation group, operators in the RHIC control room, and many other people in CAD for their help. I would like to express special thanks to Dr. Rogelio Tomás from CERN for proposing the idea of using horizontal closed orbit at sextupoles for beta-beat correction and his generous help.

I would like to express my gratitude to Yann Dutheil, Malek Haj Tahar, Zhe Duan, Yuan Hui Wu, Chaofeng Mi, Qiong Wu, Bingping Xiao, Meng Li, Hao Xu, Xiaofeng Gu,Tianmu Xing, Xue Liang, and Mengjia Gaowei for many fruitful discussions, a lot of help and fun they shared with me. I would like to thank the soccer team in BNL for a lot of joyful moments in the soccer games.

I would like to express my appreciations to my fellows in our Accelerator Physics group, Honghuan Liu, Alfonse Pham, Zhenghao Gu, Hung-Chun Chao, Micheal Ng, Alper Duru, Kilean Huang, Kun Fang, Haisheng Xu, Ao Liu, Patrick McChesney and Jeffrey Eldred, for being not only excellent colleagues but also good friends. Many thanks to all my friends who have been supporting me.

I would like to thank my Doctorate committee, Prof. Steven Gottlieb, Prof. Michael Snow, and Prof. John Carini for their insightful comments on my thesis.

Finally, I would like to thank my parents for their immeasurable love and encouragement throughout my lifetime.

The quality of beam optics is of great importance for the performance of a high energy accelerator like the Relativistic Heavy Ion Collider (RHIC). The turn-by-turn (TBT) beam position monitor (BPM) data can be used to derive beam optics. However, the accuracy of the derived beam optics is often limited by the performance and imperfections of instruments as well as measurement methods and conditions. Therefore, a robust and model-independent data analysis method is highly desired to extract noise-free information from TBT BPM data. As a robust signal-processing technique, an independent component analysis (ICA) algorithm called second order blind identification (SOBI) has been proven to be particularly efficient in extracting physical beam signals from TBT BPM data even in the presence of instrument's noise and error. We applied the SOBI ICA algorithm to RHIC during the 2013 polarized proton operation to extract accurate linear optics from TBT BPM data of AC dipole driven coherent beam oscillation. From the same data, a first systematic estimation of RHIC BPM noise performance was also obtained by the SOBI ICA algorithm, and showed a good agreement with the RHIC BPM configurations. Based on the accurate linear optics measurement, a beta-beat response matrix correction method and a scheme of using horizontal closed orbit bumps at sextupoles for arc beta-beat correction were successfully applied to reach a record-low beam optics error at RHIC. This thesis presents principles of the SOBI ICA algorithm and theory as well as experimental results of optics measurement and correction at RHIC.

Contents

List of Tables

List of Figures

Chapter 1

Introduction

The Relativistic Heavy Ion Collider (RHIC) is a high energy hadron collider located at the Brookhaven National Lab (BNL), NY, USA. Figure 1.1 shows the layout of the RHIC accelerator complex which includes the proton linear accelerator (LINAC), electron beam ion source (EBIS), a Booster ring, the Alternating Gradient Synchrotron (AGS), and the Blue and Yellow rings of RHIC. Since 2000, RHIC has been in oper-

Figure 1.1: Layout of the RHIC accelerator complex.

ation for studies involving collisions of heavy ion beams such as beams of the nuclei of gold atoms at energies up to $100\,\mathrm{GeV/u}$, and polarized protons at energies up to $255\,\mathrm{GeV}$. For heavy ion operation, RHIC has delivered high luminosity with a great operational flexibility for collisions of different species at various energies. For polarized proton operation, both luminosity and polarization are optimized to reveal the source of proton spin. Figure 1.2 shows the integrated luminosity for heavy ions and polarized protons achieved at RHIC.

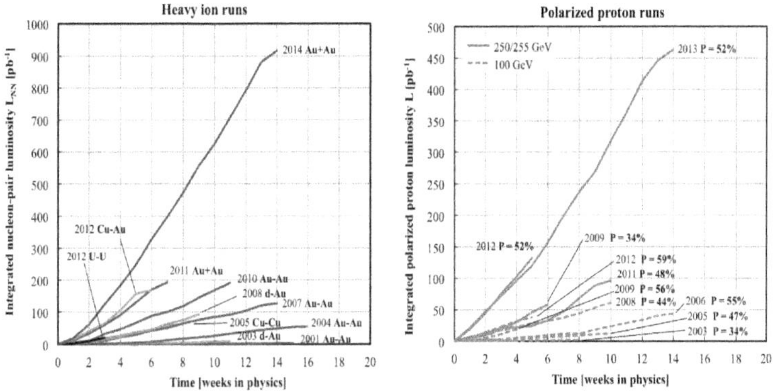

Figure 1.2: Integrated luminosity for heavy ions (left) and polarized protons (right) achieved at RHIC.

The beam optics parameters of RHIC are carefully designed to achieve the potential luminosity and polarization performance. However, large deviation of the beam optics parameters in the real machine from the design values due to imperfections in the accelerator components degrades the luminosity and polarization performance or even cause damage to the accelerator. Therefore, accurate optics measurement and efficient optics correction techniques are in great demand for routine inspection and prompt control of beam optics parameters. The main topic of this thesis is to develop such techniques for RHIC operations.

This study started in February, 2013 at the Collider-Accelerator Department of BNL. During the study, the technique of independent component analysis (ICA) was first introduced to RHIC for a systematic estimation of RHIC BPM noise performance and turn-by-turn (TBT) beam position monitor (BPM) data based optics measurement. The beta-beat response matrix based global correction scheme and the method of arc beta-beat correction using horizontal closed orbit at sextupoles were developed for RHIC. These techniques were successfully demonstrated in two beam experiments with a total of 4 hours' beam time. In the meantime, various software packages were developed to facilitate the beam experiments and future studies.

This thesis studies the optics measurement and correction technique developed for RHIC. Chapter 2 reviews fundamental of single particle beam dynamics to facilitate discussions in the rest of this thesis. Chapter 3 discusses optics measurement at RHIC. After the introduction of RHIC optics and instrumentation, various optics measurement techniques are reviewed. A time-correlation based ICA algorithm called second order blind identification (SOBI) is studied in detail. Principles and experimental results of application of SOBI for BPM noise estimation and optics measurement are discussed. Chapter 4 is devoted to optics correction at RHIC. The theories of both the beta-beat response matrix correction method and the technique of arc beta-beat correction using horizontal closed orbit at sextupoles are studied. Experimental results at RHIC are reported. Software packages developed for this study are introduced in Chapter 5.

Chapter 2

Fundamentals of single particle beam dynamics in circular accelerators

In synchrotrons, charged particles are confined in a closed orbit in a vacuum beam pipe by electromagnetic forces from magnets and radio-frequency (RF) cavities. Dipole magnets are used to bend charged particles at a given momentum along a reference circular orbit. Quadrupole magnets are used to focus charged particles. Since quadrupole magnets always exert a focusing force on one transverse direction and a defocusing force on the other transverse direction, an alternating gradient focusing scheme [5] is successfully used in all modern accelerators. The transverse oscillation of charged particles around the closed orbit is called betatron oscillation. In the longitudinal direction, charged particles interact with longitudinal electric field at the gaps of RF cavities. The arrival time of a charged particle to the gaps of RF cavities should be synchronized with the right half RF period for the charged particle to gain energy. According to the phase stability theory [3, 4], by properly setting the RF phase, more energetic particles gain less energy from the RF cavity and vice versa. Therefore, particles are bunched together in the longitudinal direction by the focusing of RF cavities.

In the remaining sections of this chapter, fundamentals of single particle beam dynamics are discussed to facilitate discussions in the following chapters. In Section 2.1, the dynamical system of a circular accelerator is described in the Frenet-Serret coordinates. Magnet fields and magnets in a circular accelerator are discussed in Section 2.2. The dynamics of transverse motion is introduced in Section 2.3, while Section 2.4 discusses the longitudinal motion. A summary is presented in Seciton 2.5.

2.1 Dynamical system in the Frenet-Serret coordinates

2.1.1 Equations of motion

The force acting on a charged particle in the electric field $\mathbf{E}$ and magnetic field $\mathbf{B}$ is the Lorentz force

$$\frac{d\mathbf{p}}{dt} = \mathbf{F} = e(\mathbf{E} + \mathbf{v} \times \mathbf{B}), \tag{2.1}$$

where $\mathbf{p} = \gamma m \mathbf{v}$ is the mechanical momentum, $\mathbf{v} = \frac{d\mathbf{r}}{dt}$ is the velocity, $\mathbf{r}$ is the displacement vector with respect to a given origin, m is the mass, e is the charge, and $\gamma = 1/\sqrt{1 - v^2/c^2}$ is the relativistic Lorentz factor. When the speed of the charged particle is close to the speed of light, i.e., $v \approx c$, the force due to the magnetic field is much stronger than the one from the electric field that can be produced in present-day technology. Therefore, modern high energy synchrotrons favor magnetic fields to guide charged particles.

The electric and magnetic fields are related to the vector potential $\mathbf{A}$ and scalar potential Φ via $\mathbf{E} = -\nabla\Phi - \partial\mathbf{A}/\partial t$, and $\mathbf{B} = \nabla \times \mathbf{A}$. Equation (2.1) can be derived from Lagrange's equation

$$\frac{d}{dt}\left(\frac{\partial L}{\partial \mathbf{v}}\right) - \frac{\partial L}{\partial \mathbf{r}} = 0, \tag{2.2}$$

where $L = -mc^2\sqrt{1 - v^2/c^2} - e\Phi + e\mathbf{v}\cdot\mathbf{A}$ is the Lagrangian. The canonical momentum $\mathbf{P}$ is

$$\mathbf{P} = \frac{\partial L}{\partial \mathbf{v}} = \mathbf{p} + e\mathbf{A}. \tag{2.3}$$

The corresponding Hamiltonian H is

$$H = \mathbf{P}\cdot\mathbf{v} - L = c\sqrt{m^2c^2 + (\mathbf{P} - e\mathbf{A})^2} + e\Phi, \tag{2.4}$$

and Hamilton's equations of motion are

$$\dot{Q}_i = \frac{\partial H}{\partial P_i}, \quad \dot{P}_i = -\frac{\partial H}{\partial Q_i}, \quad i = 1, 2, 3, \tag{2.5}$$

where the overdot is the derivative with respect to time t, and (Q_i, P_i) are three pairs of conjugate phase space coordinates with respect to any reference point.

2.1.2   Frenet-Serret coordinate system

To study particle motion in a circular accelerator, which is usually oscillations around a reference orbit, single particle beam dynamics adopts a curvilinear coordinate system called Frenet-Serret coordinate system, as shown in Fig. 2.1. The reference orbit $\mathbf{r}_0$, which is the trajectory of an ideal charged particle with nominal momentum and right initial conditions, is a closed path determined by the location and magnetic fields of the deflecting magnets. The path length s is measured along the reference orbit from an initial point. The unit vectors $(\mathbf{x}, \mathbf{s}, \mathbf{z})$ form the basis of Frenet-Serret coordinate system moving along the reference orbit $\mathbf{r}_0$, where $\mathbf{s}$ is pointing to the tangential direction at $\mathbf{r}_0(s)$, $\mathbf{x}$ is perpendicular to $\mathbf{s}$ and in the tangential plane pointing to the outer side, and $\mathbf{z} = \mathbf{x} \times \mathbf{s}$. Directions along $\mathbf{x}$ and $\mathbf{z}$ are called horizontal and vertical directions, respectively, while direction along $\mathbf{s}$ is referred to as longitudinal direction.

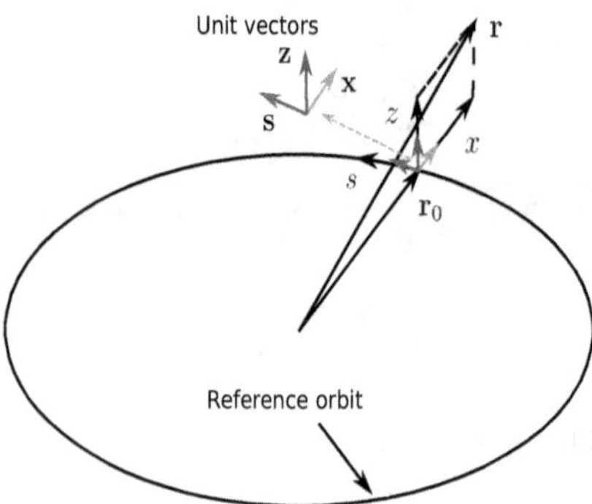

Figure 2.1: Frenet-Serret coordinate system. The unit vectors $(\mathbf{x}, \mathbf{s}, \mathbf{z})$ move along the reference orbit $\mathbf{r}_0$. Arbitrary displacement $\mathbf{r}(s) = \mathbf{r}_0(s) + x\mathbf{x}(s) + z\mathbf{z}(s)$ with s being the independent variable.

In general, particles oscillate around $\mathbf{r}_0$ in all these three directions. Because of the repetitive nature of the components of a circular accelerator in the longitudinal direction, a Hamiltonian with s as the independent variable is convenient to fruitfully exploit the physics of linear and nonlinear beam dynamics. Such a Hamiltonian is given by [6]

$$\tilde{H} = -\left(1 + \frac{x}{\rho}\right)\sqrt{\frac{(H - e\Phi)^2}{c^2} - m^2c^2 - (p_x - eA_x)^2 - (p_z - eA_z)^2} - eA_s, \qquad (2.6)$$

where ρ is the local bending radius of the reference orbit, $A_x = \mathbf{A} \cdot \mathbf{x}$, $A_s = (1 + \frac{x}{\rho})\mathbf{A} \cdot \mathbf{s}$, $A_z = \mathbf{A} \cdot \mathbf{z}$ and $(x, p_x, z, p_z, t, -H)$ are the new phase space coordinates. The energy and momentum of the particle are $E = H - e\Phi$ and $p = \sqrt{E^2/c^2 - m^2c^2}$, respectively. Since the transverse momentum p_x and p_z are much smaller than the total momentum p, Eq. 2.6 can be expanded up to second order in p_x and p_z

$$\tilde{H} = -p\left(1 + \frac{x}{\rho}\right) + \frac{1 + x/\rho}{2p}\left[(p_x - eA_x)^2 + (p_z - eA_z)^2\right] - eA_s. \qquad (2.7)$$

The quantities, $x' = dx/ds = \dot{x}/\dot{s}$, $z' = dz/ds = \dot{z}/\dot{s}$, are the deflecting angle in the horizontal and vertical direction, respectively.

2.2 Magnet fields and magnets in a circular accelerator

In a synchrotron with transverse magnetic fields, we can set $A_x = A_z = 0$ and a zero scalar potential of $\Phi = 0$. The two-dimensional magnetic field in the Frenet-Serret coordinate system is expressed as

$$\mathbf{B} = B_x(x, z)\mathbf{x} + B_z(x, z)\mathbf{z}, \tag{2.8}$$

where

$$B_x = -\frac{1}{1 + x/\rho}\frac{\partial A_s}{\partial z}, \quad B_z = \frac{1}{1 + x/\rho}\frac{\partial A_s}{\partial x}. \tag{2.9}$$

The 2D magnetic field can be expressed in a complex representation [7] with the U.S. convention as

$$B_z + iB_x = B_0 \sum_{n=0}^{\infty}(b_n + ia_n)(x + iz)^n, \tag{2.10}$$

$$b_n = \frac{1}{B_0 n!}\frac{\partial^n B_z}{\partial x^n}\bigg|_{x=z=0}, \quad a_n = \frac{1}{B_0 n!}\frac{\partial^n B_x}{\partial x^n}\bigg|_{x=z=0},$$

where i is the imaginary unit, b_n and a_n are the $2(n+1)$ multipole coefficients with dipole b_0, dipole roll a_0, quadrupole b_1, skew quadrupole a_1, sextupole b_2, skew sextupole a_2, etc[1]. The normalization constant B_0 is usually chosen as the main dipole field strength such that $b_0 = 1$.

Historically, many old circular accelerators had been built with combined-function magnets which combine a dipole field for deflection and a quadrupole field for focusing. Modern large circular accelerators usually employ separated-function magnets,

---

[1] European convention identifies b_1, a_1 for dipole and dipole roll, $b2, a2$ for quadrupole and skew quadrupole, etc.

i.e., dipole magnets for deflection, quadrupole magnets for focusing, etc. Generally speaking, separated-function magnets allow higher energy charged particles because the iron yoke of a pure dipole magnet saturates at higher field strengths than the yoke of a combined-function magnet. Moreover, the amplitude of charged particle oscillation along the horizontal direction may blow up in an accelerator built with pure combined-function magnets [8]. Therefore, the following discussions are focused on separated-function magnets.

2.2.1 Dipole

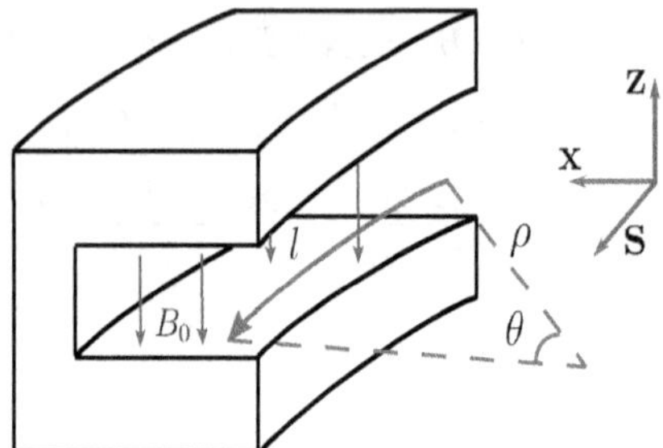

Figure 2.2: Schematic plot of a dipole magnet with field strength B_0 and trajectory of an ideal positively charged particle. Dipole length l, bending radius ρ, and bending angle θ are related by $l = \rho\theta$.

A dipole magnet provides uniform magnetic field to deflect charged particles. As shown in Fig. 2.2, the dipole length l, bending radius ρ and bending angle θ are related by $l = \rho\theta$. Given a dipole field strength B_0, the bending radius for a particle with charge e and momentum p is given by

$$\rho = \frac{p}{e}\frac{1}{B_0} = \frac{[B\rho]}{B_0}, \tag{2.11}$$

where $[B\rho] \equiv p/e$ is defined as the momentum rigidity[2].

2.2.2 Quadrupole

Quadrupole magnets are used to focus the charged particle beam in the vacuum chamber. Normal quadrupole magnetic fields depend linearly on the transverse displacement with respect to the magnet center as

$$B_x = \frac{\partial B_x}{\partial z} z, \quad B_z = \frac{\partial B_z}{\partial x} x, \tag{2.12}$$

where the magnetic field gradients in the horizontal and vertical directions are equal, i.e., $\partial B_x/\partial z = \partial B_z/\partial x = B_1$, because of the symmetry, and x, z are the horizontal and vertical displacements in the Frenet-Serret coordinate system provided that the reference orbit is beam-based aligned [9] to pass through the center of the quadrupole magnet. Figure 2.3 shows a schematic plot of a horizontal focusing normal quadrupole magnet for positively charged particles whose directions point into the paper. A horizontal focusing quadrupole magnet is defocusing in the vertical plane and vice versa. A horizontal focusing quadrupole magnet is turned into a vertical focusing quadrupole if it is rotated in the x–z plane by $90°$ and vice versa.

Skew quadrupole magnets are normal quadrupole magnets rotated by $45°$ in the x–z plane. A skew quadrupole magnet provides force in one transverse direction whose amplitude is linearly proportional to the displacement with respect to the magnet center in the other transverse plane, therefore couples the charged particle motions in both planes. Skew quadrupole magnets are usually used to compensate for the linear coupling due to the skew quadrupole magnetic field components from magnet imperfections and misalignments.

---

[2]The rectangular brackets are used to emphasize momentum rigidity $[B\rho]$ is a single physical quantity with unit $[\mathrm{T \cdot m}]$ but not simply a product.

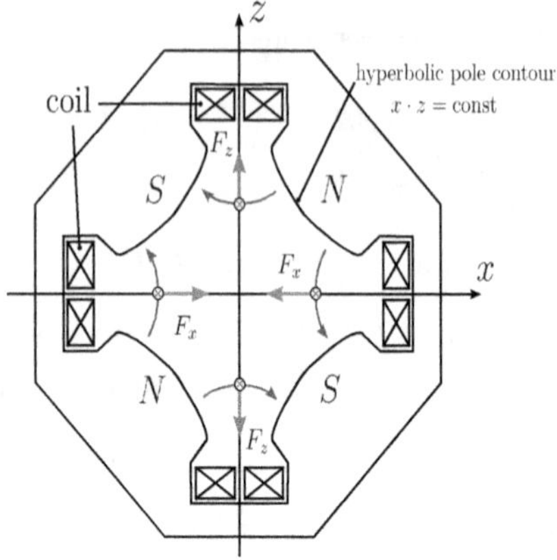

Figure 2.3: Schematic plot of a horizontal focusing normal quadrupole magnet for positively charged particles whose directions point into the paper. The blue arrows denote the magnetic field lines, the green arrows show the horizontal focusing force, and the magenta arrows show the vertical defocusing force.

2.2.3 Sextupole

Sextupole and skew sextupole magnets provide transverse magnetic fields which depend on transverse displacement to second order. The magnetic field of a normal sextupole is

$$B_x = B_2 xz, \quad B_z = B_2 \frac{x^2 - z^2}{2}, \tag{2.13}$$

where $B_2 = \partial^2 B_z / \partial x^2$ is the second order magnetic field gradient. Normal sextupoles are used to compensate for chromaticity which will be discussed in Section 2.3.6. A skew sextupole is obtained by rotating a normal sextupole by $30°$ in the x–z plane.

Aside from dipole, quadrupole, and sextupole magnets, modern circular accelerators may be equipped with some other nonlinear magnets, e.g., octupole, decatupole, etc. A sequential composition of strengths and longitudinal locations of accelerator magnets is called a magnetic lattice.

2.3   Transverse motion

The transverse oscillation of charged particles around the reference orbit in the presence of the linear magnets, i.e., dipoles and quadrupoles, is called betatron motion which is to be discussed in this section.

2.3.1 Equation of motion

Disregarding the longitudinal motion, the equations of betatron motion for a charged particle in the presence of transverse magnetic fields given by Eq. (2.7) are [6]

$$x'' - \frac{\rho + x}{\rho^2} = \frac{B_z}{[B\rho]} \frac{p_0}{p} \left(1 + \frac{x}{\rho}\right)^2, \tag{2.14}$$

$$z'' = -\frac{B_x}{[B\rho]} \frac{p_0}{p} \left(1 + \frac{x}{\rho}\right)^2, \tag{2.15}$$

where p_0 is the momentum of the ideal particle which travels along the reference orbit and $[B\rho]$ is the momentum rigidity for the on-momentum particle with $p = p_0$.

For an on-momentum particle in the presence of dipole and quadrupole magnetic fields, the betatron equations of motion in Eqs. (2.14) and (2.15) become the linear Hill's equations [10]

$$x'' + K_x(s)x = 0, \quad z'' + K_z(s)z = 0, \tag{2.16}$$
$$K_x = 1/\rho^2 - K_1(s), \quad K_z = K_1(s),$$

where $K_x(s), K_z(s)$ are the periodic focusing function due to the repetitive nature of a circular accelerator, and $K_1(s) = B_1(s)/[B\rho]$ is the normalized quadrupole field gradient.

According to Eq. (2.16), a pure sector dipole, whose entrance and exit angles are perpendicular to the edge of the dipole field, provides only horizontal focusing with $K_x = 1/\rho^2, K_z = 0$, while a quadrupole is focusing in one transverse plane but defocusing in the other transverse plane with $K_x = -K_z$.

Modern circular accelerators are usually composed of sequential identical sections, each of which is called a superperiod. The magnetic fields of each accelerator component is usually designed to be uniform or nearly uniform in the magnet main body[3]. Therefore, the focusing functions $K_x(s)$ and $K_z(s)$ are essentially piecewise constant. Let y be either x or z, Eq. (2.16) becomes

$$y'' + K_y(s)y = 0, \tag{2.17}$$

with the periodic focusing function $K_y(s + L) = K_y(s)$ where L is the length of a superperiod. The phase space coordinates (y, y') are conjugate to each other.

---

[3]In large high energy synchrotrons, the fringe fields at the edges of the magnet are usually negligible.

2.3.2   Matrix formalism

It is convenient to trace phase space coordinates (y, y') as the charged particle travels along each accelerator component for beam dynamics study. The phase space coordinates (y, y') at both ends of an element are related by a symplectic map [11], which reduces to a 2×2 transfer matrix M for a linear system such that

$$\begin{pmatrix} y \\ y' \end{pmatrix}_s = M(s, s_0) \begin{pmatrix} y \\ y' \end{pmatrix}_{s_0} \tag{2.18}$$

The transfer matrix for a few common accelerator components are listed as follows.

- Drift space

$$M_{\text{drift}} = \begin{pmatrix} 1 & l \\ 0 & 1 \end{pmatrix}. \tag{2.19}$$

- Sector dipole with bending radius ρ and deflecting angle θ

$$M_{\text{dipole}} = \begin{pmatrix} \cos\theta & \rho\sin\theta \\ -\frac{1}{\rho}\sin\theta & \cos\theta \end{pmatrix}, \tag{2.20}$$

- Quadrupole with constant focusing function K and length l

$$M_{\text{quad}} = \begin{cases} \begin{pmatrix} \cos(\sqrt{K}l) & \frac{1}{\sqrt{K}}\sin(\sqrt{K}l) \\ -\sqrt{K}\sin(\sqrt{K}l) & \cos(\sqrt{K}l) \end{pmatrix}, & K > 0, \\[2em] \begin{pmatrix} \cosh(\sqrt{|K|}l) & \frac{1}{\sqrt{|K|}}\sinh(\sqrt{|K|}l) \\ \sqrt{|K|}\sinh(\sqrt{|K|}l) & \cosh(\sqrt{|K|}l) \end{pmatrix}, & K < 0. \end{cases} \tag{2.21}$$

Modern circular accelerators are usually constructed with repetitive cells composed of a few magnets. The left plot of Fig. 2.4 shows the lattice of FODO cell, which is consist of a pair of focusing (F) and defocusing (D) quadrupoles separated by drift space (O). Replace the drift space by dipole (B), one obtains a FBDB cell which is shown in the middle plot of Fig. 2.4. The right plot in Fig. 2.4 shows the lattice of a triplet, which contains three quadrupoles with the polarity of the center quadrupole opposite to the other two quadrupoles. In a high energy circular collider, FBDB cells are frequently used for transporting charged particle beams in the arc sections, while triplets are usually used in the interaction region to strongly focus the charged particle beam into small transverse beam sizes to facilitate collisions.

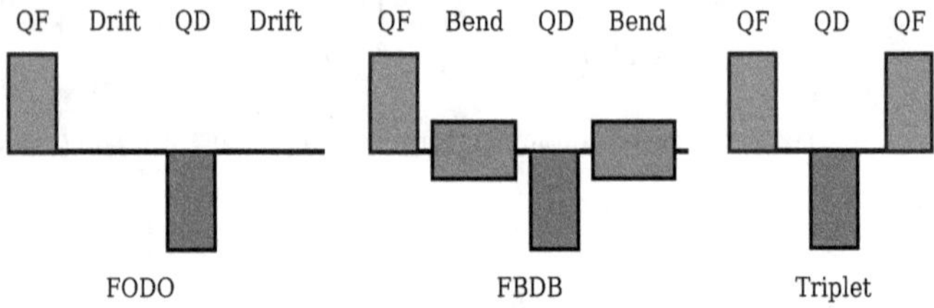

Figure 2.4: Lattice of a FODO cell (left), a FBDB cell (middle) and a triplet cell (right).

The transfer matrix of a cell of n consecutive elements is the sequential product of transfer matrices of each element

$$M(s + L, s) = M_n \cdots M_2 M_1, \tag{2.22}$$

where L is the total length of the cell.

2.3.3 Courant-Snyder parameters and beam emittance

To further explore the physics lying behind Eq. (2.17), the following forms of general solutions according to Floquet's theory [10] are useful:

$$y_1(s) = a w_y(s) e^{i\psi_y(s)}, \quad y_2(s) = a w_y(s) e^{-i\psi_y(s)}, \tag{2.23}$$

where a is a constant, w_y and ψ_y are the amplitude and phase functions, respectively. Since $K_y(s)$ is real, the amplitude and phase functions satisfy the betatron envelope and phase equations

$$w_y'' + K_y w_y - \frac{1}{w_y^3} = 0, \quad \psi_y' = \frac{1}{w_y^2}, \tag{2.24}$$

respectively[4]. By imposing periodic boundary conditions for the amplitude function[5], i.e.,

$$w_y(s) = w_y(s+L), \quad w_y'(s) = w_y'(s+L),$$

one can further introduce the Courant-Snyder parameters, or Twiss parameters,

$$\beta_y = w_y^2, \quad \alpha_y = -w_y w_y', \quad \gamma_y = \frac{1+\alpha_y^2}{\beta_y} \tag{2.25}$$

to parametrize the solutions of Hill's equations. β_y is called betatron amplitude function or beta function. With the Courant-Snyder parameters, the phase function becomes

$$\psi_y(s) = \int_0^s \frac{ds}{\beta_y(s)}. \tag{2.26}$$

An important quantity called betatron tune ν_y, defined as the number of betatron oscillations in one revolution, is

$$\nu_y = \frac{1}{2\pi} \int_s^{s+C} \frac{ds}{\beta_y(s)}, \tag{2.27}$$

where C is the circumference of the accelerator. Therefore, the betatron oscillation frequency is $\nu_y f_0$, where f_0 is the revolution frequency.

The Twiss parameters $(\beta_y, \alpha_y, \gamma_y)$ and phase function ψ_y are only determined by the accelerator lattice through the focusing function K_y. General solutions of

---

[4]The integration constant of phase equation is chosen to be zero for simplicity.

[5]Although the periodic boundary condition is not necessary, it would simplify the solution of the differential equations to aid the design of a circular accelerator.

Eq. (2.17) expressed in the Twiss parameters and phase function are

$$y = \sqrt{\epsilon_y \beta_y} \cos[\psi_y(s) + \xi_y], \tag{2.28}$$

$$y' = -\alpha_y \sqrt{\frac{\epsilon_y}{\beta_y}} \cos[\psi_y(s) + \xi_y] - \sqrt{\frac{\epsilon_y}{\beta_y}} \sin[\psi_y(s) + \xi_y], \tag{2.29}$$

where ϵ_y and ξ_y are constants determined by the initial conditions.

Using the Courant-Snyder parameters, the transfer matrix for a section can be conveniently written as

$$M(s_2, s_1) = \begin{pmatrix} \sqrt{\beta_{y,2}} & 0 \\ -\frac{\alpha_{y,2}}{\sqrt{\beta_{y,2}}} & \frac{1}{\sqrt{\beta_{y,2}}} \end{pmatrix} \begin{pmatrix} \cos \Delta\psi_y & \sin \Delta\psi_y \\ -\sin \Delta\psi_y & \cos \Delta\psi_y \end{pmatrix} \begin{pmatrix} \frac{1}{\sqrt{\beta_{y,1}}} & 0 \\ \frac{\alpha_{y,1}}{\sqrt{\beta_{y,1}}} & \sqrt{\beta_{y,1}} \end{pmatrix}, \tag{2.30}$$

where the subscript $1, 2$ denote parameters at s_1 and s_2, respectively, and $\Delta\psi_y = \psi_{y,2} - \psi_{y,1}$ is the phase advance from s_1 to s_2. The transfer matrix for a complete revolution at s can be written as

$$M(s) = \begin{pmatrix} \cos \Phi_y + \alpha_y(s) \sin \Phi_y & \beta_y(s) \sin \Phi_y \\ -\gamma(s) \sin \Phi_y & \cos \Phi_y - \alpha_y(s) \sin \Phi_y \end{pmatrix}, \tag{2.31}$$

where $\Phi_y = 2\pi\nu_y$ is the phase advance in a revolution.

The Courant-Snyder parameters and phase function at a longitudinal location s can be calculated by either numerically solving Eq. (2.24), or from the one-turn transfer matrix (2.31) [12]. Using Eq. (2.30), the Courant-Snyder parameters can be propagated to other longitudinal positions. For example, the evolution of beta function in a drift space follows a parabola

$$\beta_y(s) = \beta_y^* + \frac{(s - s_y^*)^2}{\beta_y^*}, \tag{2.32}$$

where β_y^* is also called the waist of beta function in a drift space, and s_y^* denotes the location of the beta function waist, as shown in Fig. 2.5. Another example of the evolution of phase function and beta function in a FBDB cell is shown in Fig. 2.6.

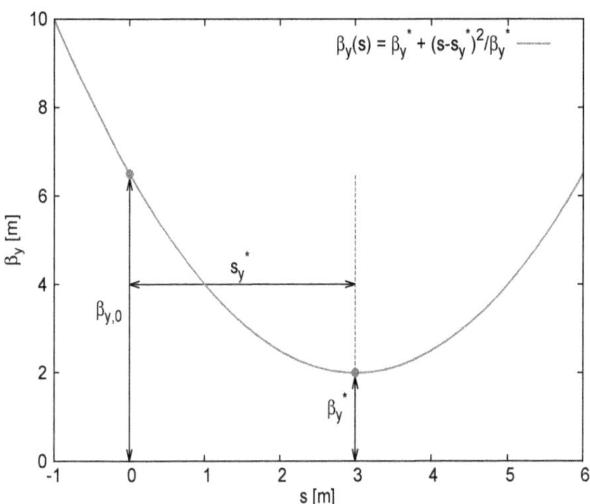

Figure 2.5: Schmatic plot of beta function in a drift space.

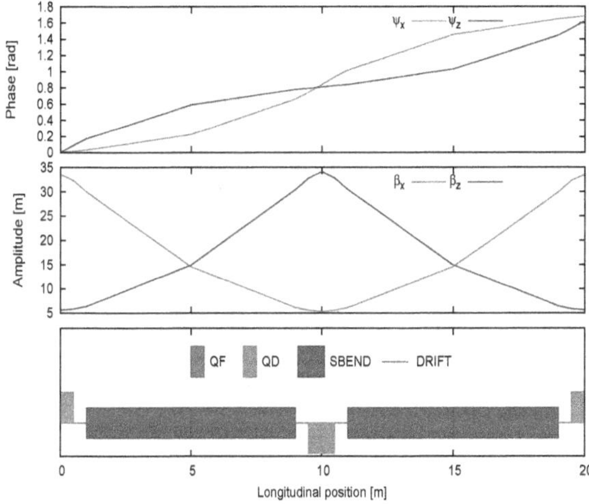

Figure 2.6: Phase function (top) and beta function (middle) for a
FBDB cell (bottom).

For this simple system, the maximum horizontal (vertical) beta function appears at the focusing (defocusing) quadrupole, while the minimum horizontal (vertical) beta function appears at the defocusing (focusing) quadrupole.

It is easy to verify from Eqs. (2.28) and (2.29) that

$$C(y, y') = \frac{1}{\beta_y}\left[y^2 + (\alpha_y y + \beta_y y')^2\right] = \gamma_y y^2 + 2\alpha_y y y' + \beta_y y'^2 = \epsilon_y \qquad (2.33)$$

is a constant called the Courant-Snyder invariant. Equation (2.33) states that the clockwise-rotation evolution of phase space coordinates (y, y') for a charged particle at a longitudinal position s trace out a Courant-Snyder ellipse whose orientation and shape are determined by the local Courant Snyder parameters $(\beta_y, \alpha_y, \gamma_y)$, as shown in Fig. 2.7. The area of the Courant-Snyder ellipse is $\pi\epsilon$. The maximum amplitude and divergence of betatron motion are $\sqrt{\beta_y\epsilon_y}$ and $\sqrt{\gamma_y\epsilon_y}$, respectively. According to

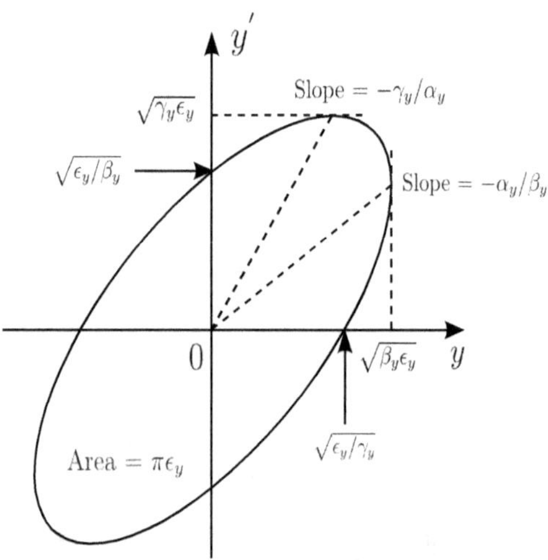

Figure 2.7: The Courant-Snyder invariant ellipse and evolution of particle phase space coordinates (y, y').

Eq.(2.33), a normalized momentum coordinate can be defined as $\mathcal{P}_y \equiv \alpha_y y + \beta_y y'$

such that the trajectory for $(y, \mathcal{P}_y)$ is a circle with a radius $\sqrt{\beta_y \epsilon_y}$.

A beam is composed of charged particles distributed in the phase space with a normalized distribution function $\rho(y, y')$ such that $\int \rho(y, y')dydy' = 1$. The moments of the beam distribution are

$$\langle y \rangle = \int y\rho(y, y')dydy', \quad \langle y' \rangle = \int y'\rho(y, y')dydy',$$

$$\sigma_y^2 = \int (y - \langle y \rangle)^2 \rho(y, y')dydy', \quad \sigma_{y'}^2 = \int (y' - \langle y' \rangle)^2 \rho(y, y')dydy',$$

$$\sigma_{yy'} = \int (y - \langle y \rangle)(y' - \langle y' \rangle)\rho(y, y')dydy' = r\sigma_y\sigma_{y'},$$

where σ_y and $\sigma_{y'}$ are the rms beam widths, $\sigma_{yy'}$ is the correlation, and r is the correlation coefficient. The rms beam emittance is defined as

$$\epsilon_{y,\mathrm{rms}} = \sqrt{\sigma_y^2 \sigma_{y'}^2 - \sigma_{yy'}^2} = \sigma_y \sigma_{y'}\sqrt{1 - r^2}. \tag{2.34}$$

Beam emittance measures the phase space area occupied by the beam to quantify the beam quality. For a Gaussian beam[6], the rms emittance ϵ_{rms} is related to the rms beam width σ_y and beta funciton β_y as[7]

$$\epsilon_{y,\mathrm{rms}} = \frac{\sigma_y^2}{\beta_y}, \tag{2.35}$$

and the emittance measuring the beam core with 95% of particles, $\epsilon_{95\%}$, is related to the rms emittance $\epsilon_{y,\mathrm{rms}}$ by

$$\epsilon_{y,95\%} = 6\epsilon_{y,\mathrm{rms}}. \tag{2.36}$$

The Courant-Snyder invariant of Eq. (2.30) derived from the phase space coordinate (y, y') is not invariant when the energy is changed. The conjugate phase space

---

[6]Ignoring dissipation and diffusion mechanisms, Gaussian distribution is commonly used for the equilibrium transverse beam distribution for an accelerator composed of linear elements such as dipoles and quadrupoles.

[7]Ignore dispersion.

coordinates (y, p_y) should be used to obtain the normalized emittance, which is the true invariant,

$$\epsilon_{n,y} = \gamma\beta\epsilon_y, \tag{2.37}$$

where γ, β are the Lorentz relativistic parameters.

2.3.4 Luminosity

A very important quantity for a collider accelerator is the luminosity $\mathcal{L}$ $[\text{cm}^{-2}\text{s}^{-1}]$ which is defined as the rate of particle collisions per unit cross-section area in a collision process. Therefore the total counting rate for a physics event is $\mathcal{R} = \sigma_{\text{phys}}\mathcal{L}$, where σ_{phys} is the cross-section of a physics process with a unit of $[\text{cm}^2]$. Since the collision point is usually designed to be at the waist of beta function in a drift space, the luminosity for two Gaussian beams is given by

$$\mathcal{L} = C\frac{f_0 N_1 N_2}{4\pi\Sigma_x\Sigma_z} \tag{2.38}$$

where C is a reduction factor due to the crossing angle, bunch length, etc., f_0 is the revolution frequency, N_1, N_2 are the intensities for beam 1 and beam 2 respectively, and

$$\Sigma_y = \sqrt{\sigma_{y1}^2 + \sigma_{y2}^2} = \sqrt{\beta_{y1}^*\epsilon_{y1} + \beta_{y2}^*\epsilon_{y2}}$$

is the convoluted beam size at the interaction point.

2.3.5 Magnets imperfections

In the presence of magnet imperfections, such as transverse or longitudinal misalignments of magnets with respect to the ideal path and magnetic field errors, Hill's equations (2.17) becomes

$$y'' + K_y(s)y = \frac{\Delta B_y}{[B\rho]}, \tag{2.39}$$

where ΔB_y is the perturbing multipole field. In this section, linear betatron motion perturbations due to dipole and quadrupole perturbing fields are discussed.

Dipole field perturbations

Dipole field imperfections may arise from errors in dipole length or power supply, dipole roll giving rise to a horizontal dipole field, a closed orbit not centered in the quadrupoles, and feed-down from higher-order multipoles. A thin dipole field error ΔB_0 in a dipole magnet at location s_0 with length ds_0 cause a kick angle $\theta = \Delta B_0 ds_0/[B\rho]$ which perturbs the closed orbit $y_{\rm co}$ at location s by

$$y_{\rm co}(s) = \frac{\sqrt{\beta_y(s)\beta_y(s_0)}}{2\sin\pi\nu_y}\cos(\pi\nu_y - |\psi_y(s) - \psi_y(s_0)|)\theta = G(s, s_0)\theta, \qquad (2.40)$$

where $G(s, s_0)$ is the Green's function of Hill's equation (2.17). From Eq. (2.40), the perturbed closed orbit oscillates around the ideal reference orbit and the number of oscillations in a revolution is near the betatron tune. Moreover, the betatron tune cannot be an integer. Otherwise, the closed orbit becomes infinity so that the motion is unstable. This is called the integer resonance.

Since Hill's equation with dipole field perturbation is linear, the closed orbit for distributed dipole field error kicks $d\theta(s_0) = \Delta B_0(s_0)ds_0/[B\rho], k = 1, 2, \cdots, N$, is a linear superposition of individual perturbations

$$\begin{aligned}
y_{\rm co}(s) &= \int_0^C G(s_k, s_0)d\theta(s_0) \\
&= \frac{\sqrt{\beta_y(s)}}{2\sin\pi\nu_y}\int_0^C \frac{\sqrt{\beta_y(s_0)}\Delta B_0(s_0)}{[B\rho]}\cos\left(\pi\nu_y - |\psi_y(s) - \psi_y(s_0)|\right)ds_0, \qquad (2.41)
\end{aligned}$$

where C is the circumference of the accelerator.

Dipole field perturbation is useful in an accelerator. In the injection region, thin dipoles, or kickers, can be used to excite local orbit bumps to facilitate beam injection. A fast kicker kicks the circulating beams out of the closed orbit for beam extraction. Correction dipoles installed along the accelerator provide capabilities of global closed orbit correction and accelerator lattice modeling via the Orbit Response Matrix (ORM) method [13].

Quadrupole field perturbations

Quadrupole field imperfections can arise from variations in the lengths of quadrupoles, errors in quadrupole power supply, horizontal closed orbit deviation in sextupoles, etc. Quadrupole field imperfections cause a perturbation in the focusing function, which which has a first-order effect on the betatron phases, tunes, and the Courant-Snyder parameters.

Consider a local quadrupole field error, or gradient error, $\Delta B_1(s_1)$ at longitudinal position s_1 with length ds_1. It perturbs the focusing function by $k(s_1)$. The change in betatron tune, or tune shift, is

$$\Delta \nu_y(s_1) = \frac{1}{4\pi} \beta_y(s_1) k(s_1) ds_1. \tag{2.42}$$

On the other hand, gradient error causes a modulation on the betatron amplitude function, or beta-beat, which is defined as

$$\frac{\Delta \beta_y(s)}{\beta_y(s)} = \frac{\tilde{\beta}_y(s) - \beta_y(s)}{\beta_y(s)}, \tag{2.43}$$

where $\tilde{\beta}_y(s)$ and $\beta_y(s)$ are the perturbed and unperturbed betatron amplitude functions, respectively. Correspondingly, there is a modulation on the phase function called phase-beat, which is defined as

$$\Delta \psi_y(s) = \tilde{\psi}_y(s) - \psi_y(s), \tag{2.44}$$

where $\tilde{\psi}_y(s)$ and $\psi_y(s)$ are the perturbed and unperturbed phase functions, respectively. For a local integrated gradient error $k(s_1)ds_1$, the resulting beta-beat is

$$\frac{\Delta \beta_y}{\beta_y}(s, s_1) = -\frac{k(s_1)\beta_y(s_1)}{2\sin(2\pi\nu_y)} \cos[2(\pi\nu_y + |\psi_y(s) - \psi_y(s_1)|)]ds_1, \tag{2.45}$$

while the corresponding phase-beat is

$$\Delta\psi_y(s, s_1) = \frac{\beta_y(s_1)k(s_1)ds_1}{4\sin(2\pi\nu_y)}\Big\{ \sin(2\pi\nu_y) + \sin\Big(2\psi_y(s_1) - 2\pi\nu_y\Big)$$
$$+ \mathrm{sign}\Big(\psi_y(s) - \psi_y(s_1)\Big)\Big(\sin(2\pi\nu_y) + \sin(2|\psi_y(s) - \psi_y(s_1)| - 2\pi\nu_y)\Big)\Big\}.$$

$$(2.46)$$

From Eqs. (2.45) and (2.46), beta-beat and phase-beat perform oscillation with about 2 times the unperturbed betatron tune. Moreover, the betatron tune cannot be a half-integer. Otherwise, the beta-beat and phase-beat become infinity so that the motion is unstable. This is called the half-integer resonance.

Figure 2.8 shows an example of phase-beat and beta-beat caused by a 5% local gradient error in a lattice composed of 8 FBDB cells. The fact that the horizontal phase-beat and beta-beat are larger indicates that the gradient error is located in a focusing quadrupole where the unperturbed β_x is larger than β_z. The location of the gradient error can also be identified by the kink at the beta-beats and the offset jump at the phase-beats. As predicted, the "tunes" for phase-beat and beta-beat are about 2 times the unperturbed betatron tunes ($\nu_x = 3.2107, \nu_z = 3.0951$) in the corresponding direction.

Due to the linearity of Hill's equation with quadrupole field perturbations, the tune shift, beta-beat, and phase-beat for distributed gradient errors are simply linear superposition of individual contributions, i.e.,

$$\Delta\nu_y = \frac{1}{4\pi}\int_0^C \beta_y(s_1)k(s_1)ds_1, \tag{2.47}$$

$$\frac{\Delta\beta_y}{\beta_y}(s) = \frac{-1}{2\sin(2\pi\nu_y)}\int_0^C k(s_1)\beta_y(s_1)\cos[2(\pi\nu_y + |\psi_y(s) - \psi_y(s_1)|)]ds_1, \tag{2.48}$$

$$\Delta\psi_y(s) = \frac{1}{4\sin(2\pi\nu_y)}\int_0^C \beta_y(s_1)k(s_1)\Big\{ \sin(2\pi\nu_y) + \sin\Big(2\psi_y(s_1) - 2\pi\nu_y\Big)$$
$$+ \mathrm{sign}\Big(\psi_y(s) - \psi_y(s_1)\Big)\Big(\sin(2\pi\nu_y) + \sin(2|\psi_y(s) - \psi_y(s_1)| - 2\pi\nu_y)\Big)\Big\}ds_1.$$

$$(2.49)$$

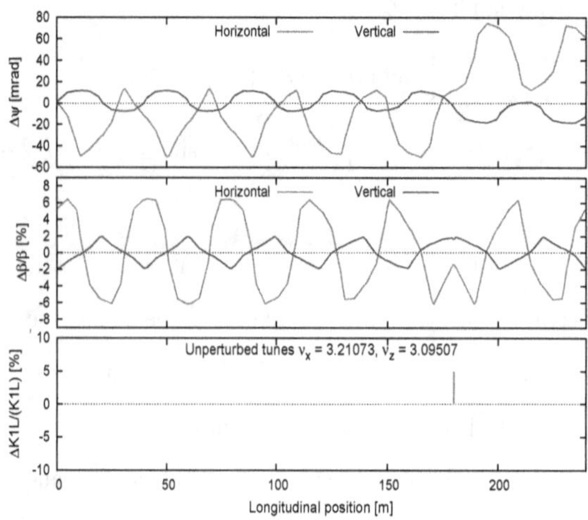

Figure 2.8: Phase-beat (top) and beta-beta (middle) for a 5% local gradient error (bottom) in a lattice with 8 FBDB cells.

Usually, gradient errors are undesired because the resulting phase-beat and beta-beat may be very harmful to beam stabilities and machine performance. However, there are still some useful applications. By scanning the gradient strength of a local quadrupole and measure the tune shifts, the average beta function at a quadrupole can be fitted according to Eq. (2.42). Recently, there is an Achromatic Telescopic Squeezing [14] (ATS) scheme which deliberately excites beta-beat to achieve smaller beta functions at the interaction point to facilitate collision without degrading beam stability.

2.3.6   Chromatic imperfections

The discussions so far involve only on-momentum particles with momentum p_0. However, a beam is composed of particles with a finite spread of momentum around p_0. To study the dynamics of an off-momentum particle with a momentum p, it is convenient to define the fractional momentum deviation as

$$\delta = \frac{\Delta p}{p} = \frac{p - p_0}{p_0}. \tag{2.50}$$

The fractional momentum deviation δ is typically small, ranging from 10^{-4} to 10^{-2}. Thus the motion of off-momentum particle can be studied perturbatively. Expanding Eqs. (2.14) and (2.15) in the presence of dipole and quadrupole fields up to first order in x, z, and δ, the equations of motion become

$$x'' + \left(K_x(s) + \Delta K_x(s) \right) x = \frac{\delta}{\rho}, \tag{2.51}$$

$$K_x(s) = \frac{1}{\rho^2} - K_1(s), \quad \Delta K_x(s) = \left(-\frac{2}{\rho^2} + K_1(s) \right) \delta \approx -K_x(s)\delta, \tag{2.52}$$

$$z'' + \left(K_z(s) + \Delta K_z(s) \right) z = 0, \tag{2.53}$$

$$K_z(s) = K_1(s), \quad \Delta K_z(s) = -K_1(s)\delta = -K_z(s)\delta. \tag{2.54}$$

The inhomogeneous term δ/ρ on the right hand side of Eq. (2.51) comes from the fact that the dipole bending angle is different for charged particles with different momentum. The effect of this inhomogeneous term δ/ρ is similar to a dipole field perturbation and gives rise to a momentum dependent orbit deviation. The perturbative focusing functions $\Delta K_x(s)$ and $\Delta K_z(s)$ can be thought of as momentum dependent gradient errors which always reduce focusing for higher momentum charged particles because they are harder to be focused.

Dispersion function

To study the chromatic effect on a horizontal closed orbit, ΔK_x in Eq. (2.51) is omitted for now, and a solution of the form

$$x = x_\beta(s) + D(s)\delta \tag{2.55}$$

can be used to obtain

$$x_\beta^{''} + K_x(s)x_\beta = 0,   \tag{2.56}$$

$$D^{''} + K_x(s)D = \frac{1}{\rho}, \quad D(s+C) = D(s),   \tag{2.57}$$

where $x_\beta(s)$ is the general solution of the homogeneous Hill's equation (2.55), which is the betatron motion, D is called the dispersion function, which is the special solution to the inhomogeneous Hill's equation (2.56), and C is the circumference of the accelerator. Treated as distributed dipole field perturbation $1/\rho(s_0)$, the dispersion function is given by

$$D(s) = \frac{\sqrt{\beta_x(s)}}{2\sin \pi\nu_x} \int_0^C \frac{\sqrt{\beta_x(s_0)}}{\rho(s_0)} \cos\left(\pi\nu_x - |\psi_x(s) - \psi_x(s_0)|\right) ds_0.   \tag{2.58}$$

Equation (2.55) states that off-momentum particles undergo horizontal betatron motion $x_\beta(s)$ around the off-momentum closed orbit given by $D(s)\delta$. The dispersion function usually vanishes in the vertical plane because the deflection is only in the horizontal direction. For a Gaussian beam, the rms beam size including the contribution from dispersion to be compared with Eq. (2.35) is

$$\sigma_y = \sqrt{\beta_y \epsilon_{y,\mathrm{rms}} + D^2\sigma_\delta^2},   \tag{2.59}$$

where σ_δ is the rms momentum deviation and the correclation between transverse and longitudinal distribution is assumed to be zero.

The total length of the closed orbit of an off-momentum charged particle can be calculated as

$$C' = \oint ds' = \oint (1 + \frac{D\delta}{\rho})ds = C + \delta \oint \frac{D(s)}{\rho(s)}ds.   \tag{2.60}$$

The momentum compaction factor is defined as fractional closed orbit length difference per unit fractional momentum deviation, i.e.,

$$\alpha_c \equiv \frac{1}{C}\frac{d\Delta C}{d\delta} = \frac{1}{C}\frac{d(C'-C)}{d\delta} = \frac{1}{C}\oint \frac{D(s)}{\rho(s)}ds.   \tag{2.61}$$

Since the revolution period $T = C/v$, the fractional difference of the revolution period of an off-momentum charged particle with respect to that of the on-momentum one is

$$\frac{\Delta T}{T} = \frac{\Delta C}{C} - \frac{\Delta v}{v} = (\alpha_c - \frac{1}{\gamma^2})\delta \equiv \eta\delta, \tag{2.62}$$

where γ is the relativistic Lorentz parameter, and η is the phase-slip factor. For a particle with the transition energy

$$\gamma_T = \sqrt{1/\alpha_c}, \tag{2.63}$$

the revolution frequency is independent of momentum deviation. The phase-slip factor and transition energy play an important role in longitudinal dynamics.

Chromaticity

The ratio of tune shift to fractional momentum deviation is called chromaticity. In particular, the chromaticity caused by the momentum dependent focusing function $\Delta K_x(s) \approx -K_x(s)\delta$ and $\Delta K_z(s) = -K_z(s)\delta$ in Eqs. (2.51) to (2.55) is called natural chromaticity, which is given by

$$C_{y,\mathrm{nat}} \equiv \frac{\Delta\nu_y}{\delta} = -\frac{1}{4\pi} \oint \beta_y K_y ds. \tag{2.64}$$

Charged particles may encounter resonance instabilities due to the tune spread caused by natural chromaticity. In modern high energy circular accelerators, sextupoles are usually used to correct natural chromaticity, since a sextupole can provide effective quadrupole focusing functions

$$\Delta K_x = S(s)D(s)\delta, \quad \Delta K_z = -S(s)D(s)\delta,$$

where $S(s) = -B_s(s)/[B\rho]$ is the effective sextupole strength and $B_s(s) = \partial^2 B_x/\partial x^2$ is the second order magnetic field gradient of the sextupole. Therefore, the chromaticity including sextupole correction is

$$C_x = -\frac{1}{4\pi} \oint \beta_x(s)[K_x(s) - S(s)D(s)]ds, \tag{2.65}$$

$$C_z = -\frac{1}{4\pi} \oint \beta_z(s)[K_x(s) + S(s)D(s)]ds. \tag{2.66}$$

On the other hand, a sextupole will introduce nonlinear perturbations to betatron motion, called geometric aberration, which produce nonlinear resonances that may endanger beam stability. The arrangement schemes have to be carefully conceived to minimize the geometric aberration [15].

2.4 Longitudinal motion

In the longitudinal direction, charged particles interact with the time-varying longitudinal electric field $\mathcal{E}$ of the RF cavity, which is given by

$$\mathcal{E}(t) = \mathcal{E}_0 \sin(h\omega_0 t + \phi_s), \tag{2.67}$$

where $\mathcal{E}_0$ is the amplitude of the electric field, h is an integer called harmonic number, $\omega_0 = \beta_0 c / R_0$ is the revolution frequency of the synchronous particle who always arrives at the RF cavity with the same phase ϕ_s called synchronous phase, $\beta_0 c$ and R_0 are the speed and average radius of the synchronous particle, respectively. Since the synchronous particle passes through the gap of the RF cavity within a finite time $t \in (nT_0 - \frac{g}{2\beta_0 c}, nT_0 + \frac{g}{2\beta_0 c})$, where g is the width of the gap, the energy gain by the synchronous particle per passage is

$$\Delta E = e\mathcal{E}_0 \beta_0 c \int_{-g/(2\beta_0 c)}^{g/(2\beta_0 c)} \sin(h\omega_0 t + \phi_s)dt = e\mathcal{E}_0 gT \sin \phi_s = eV \sin \phi_s, \tag{2.68}$$

where

$$T = \frac{\sin(hg/(2R_0))}{hg/(2R_0)}$$

is the transit time factor, and $V \equiv \mathcal{E}_0 gT$ is the effective voltage seen by the synchronous particle.

By properly choosing ϕ_s according to the phase-slip factor η, particles with fractional momentum deviation δ will oscillate around the synchronous particle in the longitudinal phase space coordinates. This longitudinal oscillation is called synchrotron motion.

2.4.1 Equations of motion and phase stability

When the acceleration $\dot{E} = \omega_0 \Delta E / 2\pi$ is low, the synchrotron motion is described by the differential equations [6]

$$\dot{\phi} = h\omega_0 \eta \delta, \tag{2.69}$$

$$\dot{\delta} = \frac{\omega_0}{2\pi \beta^2 E} eV(\sin\phi - \sin\phi_s), \tag{2.70}$$

where the dot indicates derivative with respect to time t, and the longitudinal phase space coordinates (ϕ, δ) are the the RF phase and fractional momentum deviation δ of a charged particle, respectively.

Equations (2.69) and (2.70) can be combined into a second order differential equation

$$\ddot{\phi} - \frac{h\omega_0^2 eV\eta}{2\pi\beta^2 E}(\sin\phi - \sin\phi_s) = 0. \tag{2.71}$$

When the oscillation amplitude $|\phi - \phi_s|$ is small, Eq. (2.71) can be linearized as

$$\ddot{\phi} - \frac{h\omega_0^2 eV\eta \cos\phi_s}{2\pi\beta^2 E}(\phi - \phi_s) = 0, \tag{2.72}$$

which has the form of a differential equation for a simple harmonic oscillator. The stability condition for Eq. (2.72) is $\eta \cos\phi_s < 0$ [3,4]. Therefore, below the transition energy with $\gamma < \gamma_T$ or $\eta < 0$, the synchronous phase should be $0 < \phi_s < \pi/2$. Physically, this is because the revolution period of a charged particle with fractional momentum deviation $\delta > 0$ below transition is shorter than the synchronous particle according to Eq. (2.62), i.e., it arrives earlier than the synchronous particle to the RF cavity. Thus when $0 < \phi_s < \pi/2$, particles with $\delta > 0$ pick up less energy than the synchronous particle, and vice versa, as described by Fig. 2.9. Similarly, above the transition energy with $\gamma > \gamma_T$ or $\eta > 0$, $\pi/2 < \phi_s < \pi$. The angular synchrotron frequency ω_s and synchrotron tune ν_s of Eq. (2.72) are

$$\omega_s = \omega_s \sqrt{\frac{heV|\eta \cos \phi_s|}{2\pi \beta^2 E}}, \tag{2.73}$$

$$\nu_s = \frac{\omega_s}{\omega_0} = \sqrt{\frac{heV|\eta \cos \phi_s|}{2\pi \beta^2 E}}. \tag{2.74}$$

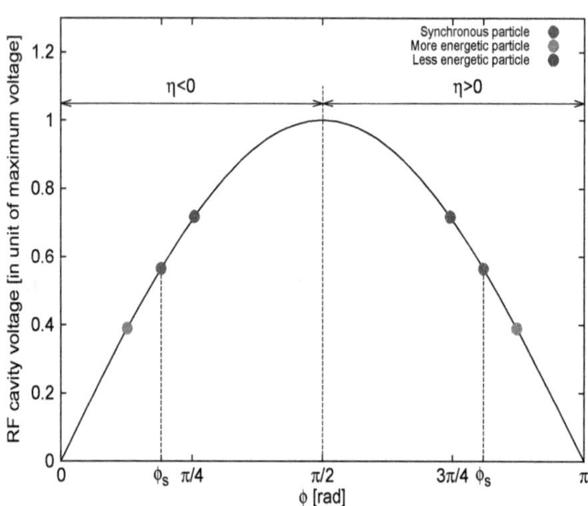

Figure 2.9: Schematic plot to illustrate phase stability theory [3,4]. Below transition $\eta < 0$, $0 \leq \phi_s \leq \pi/2$, while above transition $\eta > 0$, $\pi/2 \leq \phi_s \leq \pi$.

2.4.2 RF bucket and longitudinal emittance

The equations of motion (2.69) and (2.70) can be derived from a Hamiltonian

$$H = \frac{1}{2}h\omega_0\eta\delta^2 + \frac{\omega_0 eV}{2\pi\beta^2 E}\left[\cos\phi - \cos\phi_s + (\phi - \phi_s)\sin\phi_s\right]. \tag{2.75}$$

Equation (2.75) is essentially equivalent to a Hamiltonian for a one dimensional simple pendulum which has two fixed points $(\phi_s, 0)$ and $(\pi - \phi_s, 0)$ where $\dot{\phi} = \dot{\delta} = 0$. This Hamiltonian is a constant of motion because it is time-independent. Particles follow curves of constant Hamiltonian value H called Hamiltonian tori depending on their initial phase space coordinates. Small amplitude phase space trajectories around $(\phi_s, 0)$, the stable fixed point (SFP), are ellipses, while the phase space trajectories near $\pi - \phi_s, 0$, the unstable fixed point (UFP), are hyperbola. The Hamiltonian torus which passes through the UFP is called the separatrix, which divides the synchrotron phase space into stable and unstable regions. Figure 2.10 shows an example of separatrices for $\eta < 0$ with $\phi_s = 0, \pi/6, \pi/3$. The tori inside the separatrix are closed and bounded so the motion is stable, while the torus outside the separatrix are open so that particles outside the separatrix are not synchronized with the RF and will be lost eventually. The phase space area enclosed by the separatrix is called the bucket area. The maximum momentum deviation of the separatrix is called the bucket height. Particles inside the RF bucket are grouped together by the RF bucket to form a bunch. Since there are altogether h RF buckets for a single RF cavity, the harmonic number determines the theoretical maximum number of bunches in a circular accelerator.

In an RF bucket, particles are typically populated in the synchrotron phase space in the vicinity of the SFP. For small amplitude synchrotron motion, the equilibrium distribution is usually modeled as Gaussian. Then the phase space area the beam

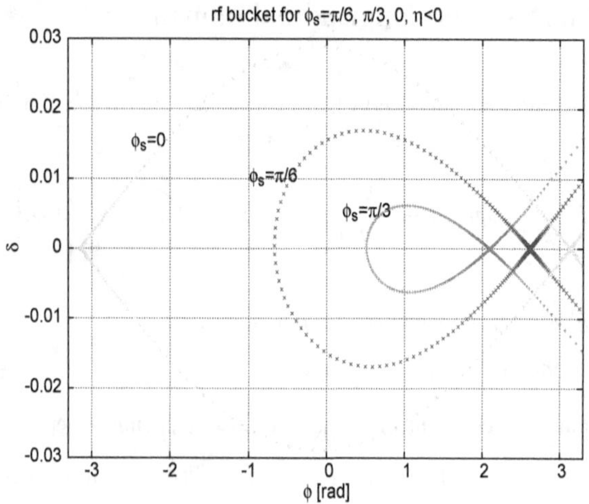

Figure 2.10: Separatrices for $\eta < 0$ with $\phi_s = 0, \pi/6, \pi/3$.

occupies is related to the rms momentum spread σ_δ and rms bunch length σ_ϕ by

$$\tilde{A}_{\mathrm{rms}} = \pi \sigma_\delta \sigma_\phi, \quad \tilde{A}_{0.95} = 6\tilde{A}_{\mathrm{rms}}, \tag{2.76}$$

where $\tilde{A}_{\mathrm{rms}}$ is the rms phase space area in (ϕ, δ), and $\tilde{A}_{0.95}$ measures 95% of the phase space area occupied by the whole bunch. The synchrotron phase space area of a bunch is usually measured in units of eV·s in the $(\phi/h, \Delta E/\omega_0)$ phase space and is related to $\tilde{A}$ by

$$A = \frac{\beta^2 E}{h\omega_0} \tilde{A}. \tag{2.77}$$

The synchrotron phase space area, which is also called the longitudinal emittance, is an important measure of beam quality.

Notice when the acceleration rate is high, the Hamiltonian tori in phase space (ϕ, δ) are not closed curves. Figure 2.11 shows two tori in (ϕ, δ) during acceleration with parameters $V = 100\mathrm{kV}, h = 1, \alpha_c = 0.04340, \phi_s = \pi/6$ at 45 MeV proton kinetic energy. The RF bucket area is shrinking during the acceleration, which is a longitudinal analog of the adiabatic damping of transverse phase space area discussed in Section 2.3.3.

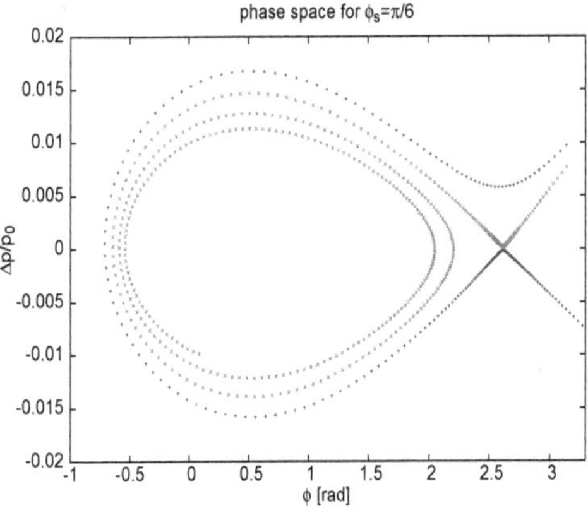

Figure 2.11: Two tori in phase space coordinates (ϕ, δ) to illustrate longitudinal adiabatic damping.

2.5   Summary

In this chapter, the fundamentals of single particle beam dynamics for both the transverse and longitudinal directions are briefly reviewed. The discussions of betatron and synchrotron motions aid the design and performance improvement of a circular accelerator.

For the purpose of this thesis, charged particle motions in the transverse and longitudinal directions are discussed separately to simplify the description of the dynamics of betatron motion and synchrotron motion, respectively. However, betatron and synchrotron motions are coupled together in general and deserve a study which treats synchrotron and betatron motion on an equal footing [16]. Besides single particle beam dynamics, a more complete framework of beam dynamics in an accelerator includes the collective beam dynamics in which the interaction between charged particles through space charge effect [17,18], synchrotron radiation [6,19,20] and conducting beam pipe as well as other relevant accelerator components [21] are studied. In addition, dynamics involving scattering, such as intrabeam scattering [22,23] and Touschek scattering [24], are also important topics for accelerator physics. However, studies of synchro-betatron coupling, collective beam dynamics and beam scattering effects are beyond the scope of this thesis so that they will not be addressed in this study. Readers interested in these topics may refer to the corresponding references cited above for more details.

Chapter 3

Optics measurement at RHIC

Since the study of charged particles transport with electromagnetic fields is analogous to the study of light transport with lenses, single particle beam dynamics is also called beam optics. Beam optics parameters such as the Courant-Snyder parameters, tune, phase advance, and chromaticity, are among the most important properties of a synchrotron, because they affect the luminosity of a collider synchrotron or the brightness of a light source synchrotron, polarization of charged particle beam, beam life time, etc. Therefore, the accelerator lattice has to be carefully designed and optimized for beam optics in order to achieve the best performance. Computer codes such as MAD [25] and ELEGANT [26] are usually used to aid the lattice design and produce a lattice model according to which the accelerator is constructed. However, beam optics parameters in the real machine will deviate from those of the lattice model due to various errors introduced during the construction and operation, which may cause a significant degradation of machine performance. Therefore, measurement and control of beam optics parameters of the real machine are highly desired for accelerator operations.

The turn-by-turn beam position monitor (TBT BPM) data can be used to extract beam optics parameters. The accuracy of the extracted beam optics is determined by the performance limitations and measurement conditions of beam instruments as well as the method of the data analysis. A robust and model-independent data analysis method is of great importance to retrieve rich information from TBT BPM data. As a robust signal-processing technique, independent component analysis (ICA) has been proven to be particularly efficient in extracting physical beam signals from TBT BPM data for beam optics measurements [27–30]. In the 2013 RHIC polarized proton operation, ICA was first applied to the Relativistic Heavy Ion Collider (RHIC) for systematic estimation of BPM noise performance from TBT BPM data of AC dipole driven beam coherent oscillation. A good agreement was found between the BPM noise estimation and the configuration of RHIC BPMs. Linear beam optics was also extracted from the same TBT BPM data and a large beta-beat was discovered in the baseline machine optics.

This chapter is dedicated to beam optics measurements at RHIC. The designed optics of RHIC is introduced in Section 3.1. A brief review of RHIC instrumentation is presented in Section 3.2. Section 3.3 reviews selected optics measurement techniques. Theoretical bases and experimental results of ICA for TBT BPM data analysis, including estimation of BPM noise performance as well as optics measurements, are given in Section 3.4 .

3.1　Introduction to RHIC optics

RHIC is composed of two identical non-circular concentric rings, which are called the Blue ring and the Yellow ring, in a common horizontal plane in the RHIC tunnel with a circumference of 3833.85 m. The two rings are oriented to intersect with one another at six crossing points called interaction points (IPs) as shown in Fig. 3.1. The IPs are labeled according to the clock manner, e.g., the southern IP is called IP6, the

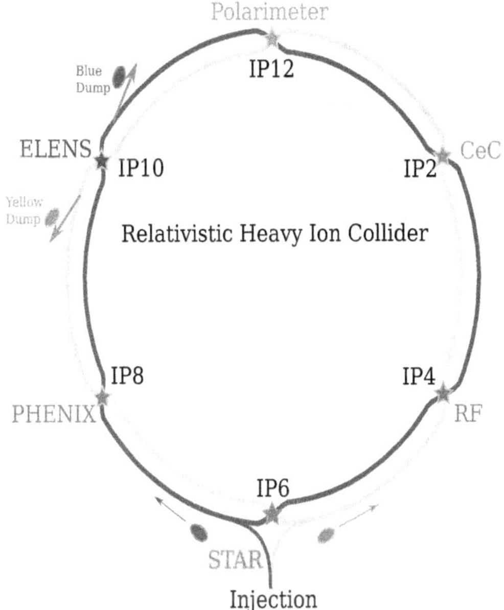

Figure 3.1: Schematic plot of the layout of RHIC.

northern IP is called IP12, etc. Each IP is located at the center of a section called insertion which brings the separated beam pipes of the two rings into an interaction region common to both rings. The section in between two insertions is called an arc that transports charged particle beams from one insertion to another. Each ring consists of three inner arcs, three outer arcs and six insertions joining the inner and outer arcs.

Table 3.1: General beam parameters for RHIC [1]

Element	Proton	Deuterium	Copper	Gold
Atomic number Z	1	1	29	79
Mass number A	1	2	63	197
Rest energy [GeV/u]	0.93827	0.93781	0.92022	0.93113
Injection:				
Kinetic energy [GeV/u]	28.3	13.7	12.6	10.8
Lorentz γ	31.2	15.6	14.5	12.6
Normalized emittance ϵ_n [mm·mrad]	20	10	10	10
Bunch area [eV·s]	0.5	0.5	0.5	0.5
RMS Bunch length [m]	2.58	4.1	4.6	5.62
Energy spread [$\times 10^{-3}$]	1.26	1.63	1.59	1.49
No. ions/Bunch [$\times 10^{9}$]	100	100	2.7	1.0
Top energy:				
Kinetic energy [GeV/u]	250.7	124.9	114.9	100.0
Lorentz γ	268.2	134.2	124.5	108.4
RMS bunch length [m]	0.10	0.17	0.18	0.19
Energy spread [$\times 10^{-3}$]	0.83	1.35	1.41	1.49

Protons and other heavy ions are injected from the insertion at IP6 into both rings and accelerated. Charged particle beams in the Blue ring circulate in clockwise direction, while those in the Yellow ring in anti-clockwise direction. General beam parameters of selected elements for RHIC are listed in Table 3.1.

The beam dump systems for both rings are located in the insertion at IP10. There are two detectors, STAR at IP6 and PHENIX at IP8. The RF cavities are installed in the insertion at IP4. The polarimeter [31] for polarization measurement is housed in the insertion at IP12. The electron lenses (ELENS) [32] for beam-beam effect compensation are being commissioned at IP10. IP2 is reserved for a proof-of-principle experiment of the coherent electron cooling (CeC) [33] in RHIC 2014 operation.

An arc of each ring consists of 11 identical FBDB cells. The main parameters and optics for an arc FBDB cell are shown in Table 3.2 and Fig. 3.2, respectively. The

Table 3.2: Main parameters of the arc FBDB cell [1]

	Inner arc FBDB	Outer arc FBDB
Length L [m]	29.5871	29.6571
Deflecting angle θ [mrad]	77.8481	77.8481
Average radius of curvature ρ [m]	380.0443	380.9443
Phase advance $(\Delta\psi_x, \Delta\psi_z)$ $[2\pi]$	(0.2234, 0.2369)	(0.2241, 0.2376)
Beta function $(\beta_{x,\mathrm{max}}, \beta_{x,\mathrm{min}})$ [m]	(49.71, 10.56)	(49.84, 10.53)
Beta function $(\beta_{z,\mathrm{max}}, \beta_{z,\mathrm{min}})$ [m]	(48.55, 9.82)	(48.70, 9.79)
Dispersion function $(D_{\mathrm{max}}, D_{\mathrm{min}})$ [m]	(1.841, 0.939)	(1.837, 0.936)
Chromaticity $(C_{x,\mathrm{nat}}, C_{z,\mathrm{nat}})$	(-0.273 -0.284)	(-0.275, -0.286)

insertions transport charged particle beams from one arc to another and control the lattice parameters at the IP's. Figure 3.3 shows a schematic plot of the layout of a

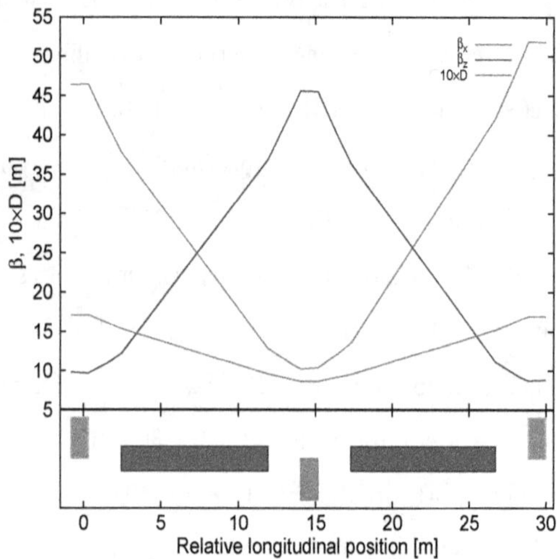

Figure 3.2: Optics of an arc FBDB cell.

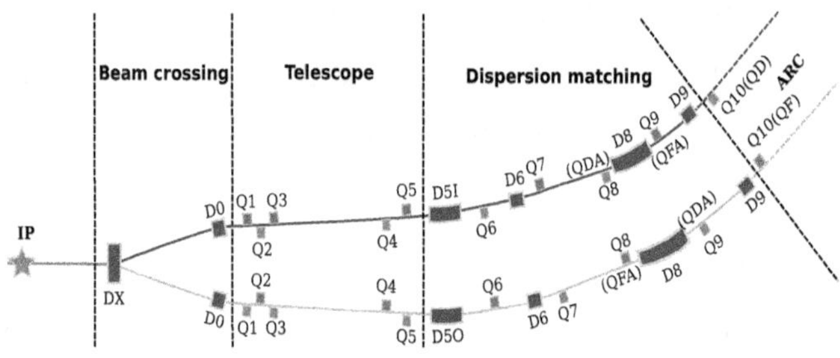

Figure 3.3: Schematic layout of a half insertion.

half of an insertion which is composed of a beam crossing section with dipoles DX and D0, a beta function matching section ("telescope") with the quadrupole triplet Q1, Q2, and Q3 as well as the quadrupole doublet Q4 and Q5, and a dispersion matching section with dipoles D5, D6, and D9 as well as quadrupoles Q6, Q7, Q8, and Q9. The large aperture dipole DX is common to both rings, but electrically and cryogenically connected to the Blue ring. The magnets Q1,Q2,Q3, and D0 of inner and outer insertions sit in common vacuum vessels. D5 of the inner and outer insertions serve also to bring the beam-beam separation from 90 cm to 41.5 cm at the edge of D0. The magnets D6 and D9 are identical, and D8 are the same as arc dipoles. Q9 of the inner insertion is identical to Q8 of the outer insertion and they are also denoted as QFA in lattice calculations. Similarly, Q8 of the inner insertion and Q9 of the outer insertion are also referred to as QDA. There is near perfect reflection anti-symmetry relative to the IP in component location and component strength.

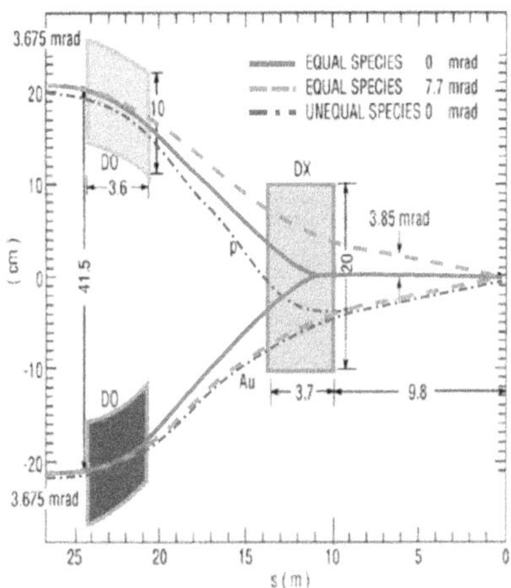

Figure 3.4: Beam crossing geometry [1].

Figure 3.4 shows the geometry of the beam crossing section, where DX is common to both beams and D0 of the inner and outer insertions are separately excited to accommodate variations in beam crossing angles as well as collision between unequal species. The crossing angle varies from 0 to 7.7 mrad. The beta function matching section serves to vary the β^* at the IP's from 0.7 to 10 m. The high beta insertion with $\beta^* = 10$ m is for injection and non-colliding IP's, while the low beta insertion with $\beta^* = 0.7$ m is for colliding IP's [1] to achieve higher luminosity. The dispersion matching section matches the optics at the exit of an insertion with the adjacent arc. Figure 3.5 shows an example of the optics for the high and low beta insertions for the Au-Au 2014 lattice.

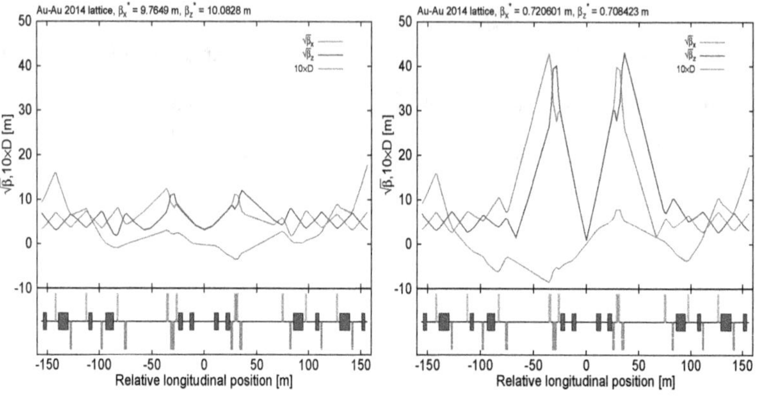

Figure 3.5: Optics of high (left) and low (right) beta insertions for the Au-Au 2014 lattice.

For RHIC, possible operating tunes, or working points, range from $\nu_{x,z} = 27.82$ to 29.20. The tune of the machine can be changed by varying the phase advance in the arc cells as well as the focusing strength of the insertion quadrupoles while maintaining the matched conditions. It has been demonstrated that the insertion is tunable over

---

[1] In the Au-Au 2014 operation, $\beta^* \approx 0.5$ m was demonstrated for colliding IP's.

Table 3.3: Available working points for RHIC [1]

	ν_x	ν_z
Equal tune	28.83	28.82
Half integer tune	28.56	28.55
Split tune	28.83	27.82
Split tune	29.20	28.20
Alternate tune	28.20	28.20
Integer tune	28.96	28.95
Nominal tune	28.19	29.18

the full β^* range at the working points shown in Table 3.5. A better working point for polarized proton at top energy was found to be around $(\nu_x, \nu_z) = (28.690, 29.685)$ [34].

RHIC is also equipped with magnets for lattice corrections, such as dipole correctors for closed orbit correction, quadrupoles for γ_T-jump [35] and linear optics correction, skew quadrupoles for linear coupling and tune splitting correction, sextupoles for chromaticity corrections, and octupoles, decapoles as well as dodecapoles for nonlinear optics correction. Table 3.4 summarizes type and number of selected magnets in both rings of RHIC, where the term "corrector" is known as dipole corrector without further explanation.

3.2 Brief review of RHIC instrumentation

RHIC is equipped with comprehensive beam diagnostics and manipulation instruments to guarantee the machine operation. In this section, the most important beam

Table 3.4: RHIC magnet inventory

Standard aperture (8 cm) arc components	
Dipoles	264
Quadrupoles	276
Sextupoles	276
Correctors	276
Insertion components	
Standard aperture (8 cm) magnets	
Dipoles (D5I, D5O, D6, D8, D9)	96
Quadruples (Q4-Q9)	144
Trim quadrupoles (@ Q4, Q5, Q6)	72
Sextupoles @ Q9	12
Correctors	144
Intermediate aperture (10 cm) magnets	
Dipoles (D0)	24
Helical dipoles	12
Intermediate aperture (13 cm) magnets	
Quadrupoles (Q1-Q3)	72
Correctors	72
Large aperture (18 cm) magnets	
Dipoles (DX)	12
Totals	
Dipoles	408
Quadrupoles	492
Trim quadrupoles	72
Sextupoles	288
Correctors	492
Total magnets	1752

Table 3.5: Mechanical details of RHIC BPM

BPM type	Warm/Cold	Plates	Locations	Aperture [cm]
1	cold	Single	Q5, Q6, Q9, arc	6.9088
2	cold	Dual	Q4, Q7, Q8	6.9088
3	cold	Dual	Q1, Q3,	11.2725
4	warm	Dual	DX, Dump, other warm	11.2725

instruments are introduced.

Beam position monitor

The beam position monitor (BPM) measures the beam centroid positions in the transverse directions. Single-plane BPMs can measure beam centroid only in one transverse direction, while dual-plane BPMs is capable of recording in both transverse directions. At RHIC, there are 176 single-plane BPMs distributed in the locations of the arcs where the beta function in the corresponding plane is maximum, and 72 dual-plane BPMs in the six insertions. There are 4 types of BPMs categorized by the their mechanical details which are listed in Table 3.5. The BPM electronics enable a 1 μm position resolution over a ±32 mm measurement range [36] and the capture of beam position within 1024 or 4096 consecutive turns. The closed orbit can also be measured by averaging the turn-by-turn beam position data from a BPM. The RHIC BPM calibration error is estimated to be about 1.0% [37].

Beam loss monitor

The purpose of the beam loss monitor (BLM) system is to prevent the quenching of RHIC magnets due to beam loss, provide quantitative loss data for tuning and archiving, and the loss history in the event of a beam abort. The RHIC BLM system uses 429 ion chambers to detect photons caused by local beam loss. The BLM system is capable of detecting fast (single turn) and slow (100 ms) loss. If the loss level exceeds the threshold for the superconducting magnets, the beam will be aborted automatically. Figure 3.6 shows a screen shot of the RHIC BLM data visualization application. The top graph shows the time evolution of raw beam loss data at each BLM, the middle graph shows the average loss rate at each BLM, and the bottom graph shows the lattice and location of the BLMs. The lattice starts from IP6 going through the Blue ring in a clockwise manner. Large beam loss was observed at collimators near IP8.

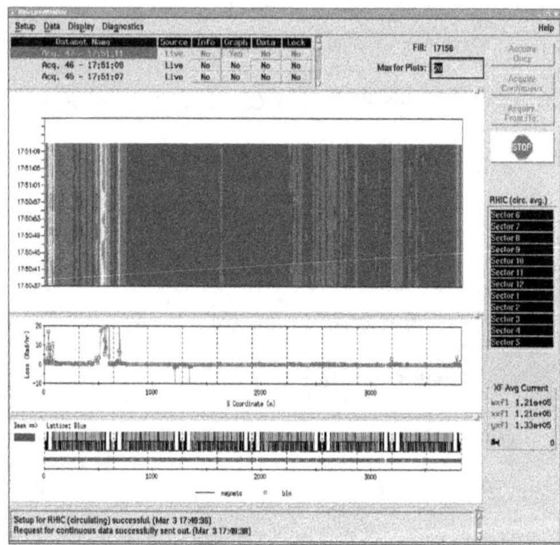

Figure 3.6: Screen shot of RHIC BLM data visualization application showing raw loss data over time (top), loss rate (middle), and lattice as well as location of BLMs (bottom).

Beam current monitor

The beam current monitor system utilizes one direct current current transformer (DCCT) for each ring and corresponding electronics to provide high resolution long term beam decay rate monitoring and low resolution short term measurement for data logging. Figure 3.7 shows the evolution of beam loss rate and intensity for a fill during the Au-Au 2014 operation measured by DCCT.

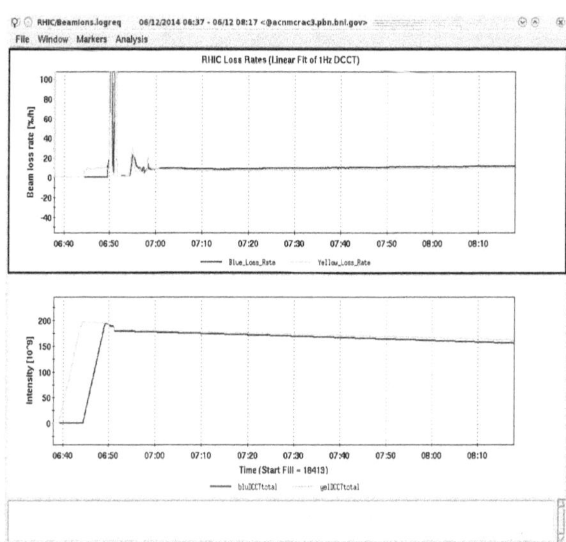

Figure 3.7: Evolution of beam loss rate (top) and intensity (bottom) from DCCT.

Wall current monitor

At RHIC, two wall current monitors (WCM's) are installed for longitudinal beam shape measurement, including bucket fill pattern measurement and bunch parameters calculation such as bunch length and peak current. Figure 3.8 shows an example of the bunch fill pattern in all RF buckets measured by WCM for the 2014 Au-Au operation, where the negative intensities of Yellow bunches are solely for a better display and should have been taken the positive values for interpretation. There is an abort gap with a few empty RF buckets to facilitate the beam abort system, as shown in Fig. 3.8 for case of the Blue ring. Notice the abort gap in the Yellow ring was filled by the leakage of gold particles from other buckets. A gap cleaning machineary in the beam dump system will have to be activated to remove the beam leaking into the abort gap.

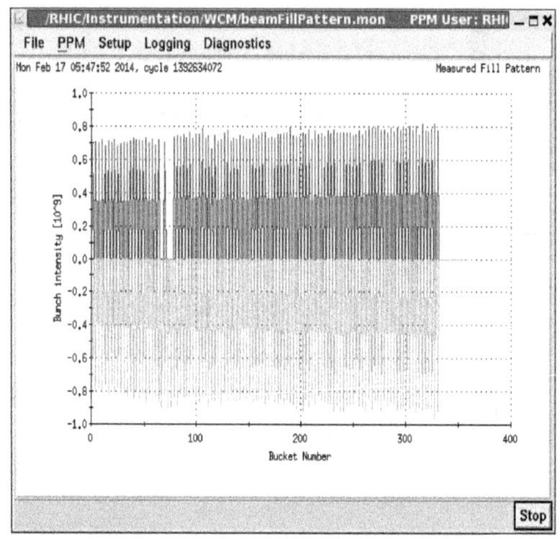

Figure 3.8: Bunch fill pattern in all RF buckets measured by the WCMs.

Ionization profile monitor

The ionization profile monitor (IPM) collects electrons that are produced as a result of beam-gas interactions for beam transverse profile measurement [38]. With the measured transverse beam size, transverse emittance can be derived with a knowledge of local beta functions and dispersion, according to Eq. (2.59). Figure 3.9 shows a screen shot of RHIC IPM application for a beam profile measurement.

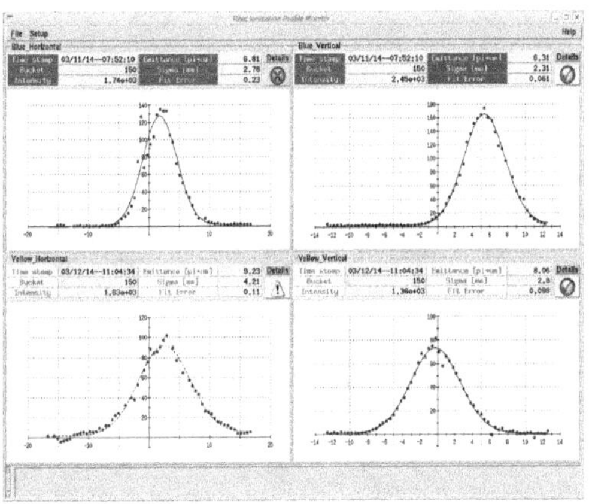

Figure 3.9: Screen shot of RHIC IPM application.

Tunemeter

At RHIC, the tune measurement device, A Rhic TUne measurement System (AR-TUS) [39], consists of fast horizontal and vertical kicker magnets and one-dual plane BPM per ring. To measure the machine tunes, coherent beam betatron oscillation are excited with a fast transverse kicker magnet [40] and transverse beam position are recorded by the BPM. The fractional tunes are derived from the position data by performing a fast Fourier transform (FFT) analysis.

The coherent beam betatron oscillation excited by ARTUS kickers is also called free betatron oscillation. TBT data of beam centroid of free betatron oscillation can be acquired at all BPMs in each ring for beam optics measurement. For free beta-tron oscillation, there are betatron tune spreads among the charged particles because

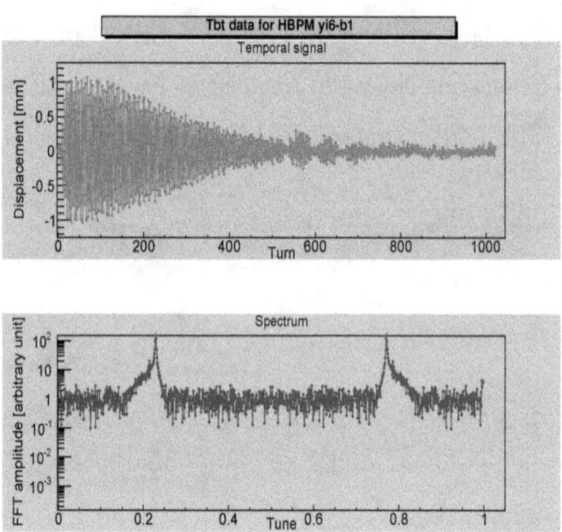

Figure 3.10: Example of free betatron oscillation TBT BPM data (top) and its FFT spectrum (bottom) recorded at the vertical BPM "yi6_b1".

of chromaticity and the effects of nonlinear multipole elements in the accelerator. Therefore, the recorded TBT BPM data will be decohered due to the accumulated betatron phase spread between charged particles [41, 42], as shown in Fig. 3.10. As a result, the free betatron oscillation TBT BPM data is prone to quick decoherence which compromises the quality of the data. In addition, the ARTUS kicker excitation can cause beam emittance growth. For high energy beam with large beam rigidity, the kicker strength may not be sufficient to excite useful free betatron oscillation with a fairly large amplitude.

AC dipole

RHIC is also equipped with AC dipoles [43] which are capable of providing time-

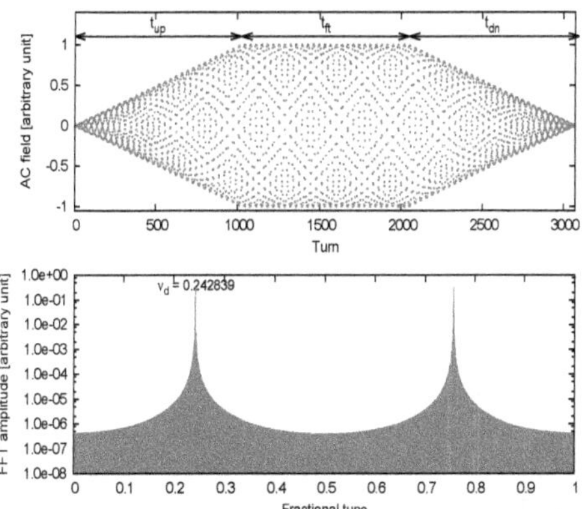

Figure 3.11: Schematic plots of AC dipole field variation over time (top) in a complete operation period and its FFT spectrum (bottom).

varying dipole fields

$$B(t) = B_m(t)\cos(2\pi\nu_d t + \psi_d), \tag{3.1}$$

where $B_m(t)$ is the amplitude profile and ν_d is the AC dipole tune. The amplitude profile $B_m(t)$ is usually adiabatically varied over time as

$$B_m(t) = \begin{cases} B_{m,0}\frac{t}{t_{\mathrm{up}}}, & 0 < t < t_{\mathrm{up}}, \\[2mm] B_{m,0}, & t_{\mathrm{up}} < t < t_{\mathrm{up}} + t_{\mathrm{ft}}, \\[2mm] B_{m,0}\left(1 - \frac{t - t_{\mathrm{up}} - t_{\mathrm{ft}}}{t_{\mathrm{dn}}}\right), & t_{\mathrm{up}} + t_{\mathrm{ft}} < t < t_{\mathrm{up}} + t_{\mathrm{ft}} + t_{\mathrm{dn}}, \end{cases}$$

where t_{up} is the time for the amplitude to be linearly ramped up to the desired field $B_{m,0}$, t_{ft} is the time when the amplitude is kept constant, and t_{dw} is the time to linearly ramp the amplitude down to zero. Figure 3.11 shows a schematic plot of the AC dipole field variation over time in a complete operation period. Such an AC dipole is able to excite a sustained coherent driven betatron oscillation of the

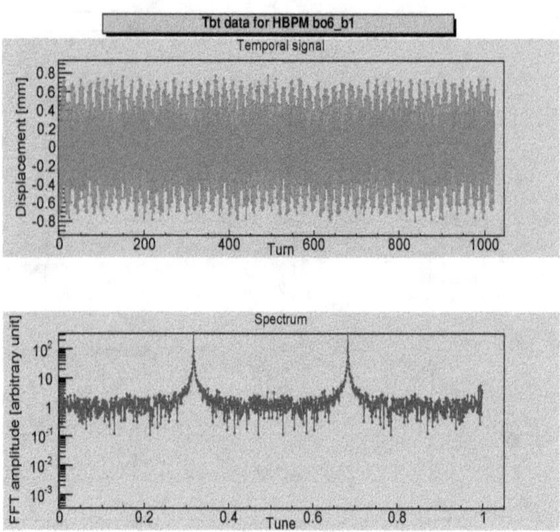

Figure 3.12: Example of driven betatron oscillation TBT BPM data (top) and its FFT spectrum (bottom) recorded at the vertical BPM "bo6_b1".

beam with a large amplitude and preservation of beam emittance when operating in an adiabatic manner [44]. An example of TBT BPM data for driven oscillation is shown in Fig. 3.12. In addition, AC dipole also efficiently excites high energy beams. Therefore, the higher signal to noise ratio as well as nondestructive nature makes AC dipole a preferable diagnostic tool for high energy accelerators.

Miscellaneous

Other important beam instrumentation at RHIC includes simultaneous orbit, tune, coupling and chromaticity feedback [45], zero degree calorimeter (ZDC) for luminosity measurement [46], collimators for limiting aperture to prevent damage due to beam loss [47], polarimeter for polarization measurement [31], stochastic cooling to reduce beam emittance [48], etc.

Many dedicated applications and advanced software are developed to facilitate the control of beam instruments as well as acquisition of beam data. Beam instrument status and machine operation data are routinely logged in the RHIC database. The beam instrumentation at RHIC has been playing an indispensable role in guiding the successful machine operation for the last decade.

3.3 Brief review of optics measurement techniques

There are various techniques using different instruments and algorithms available for optics measurement in a synchrotron. In this section, a brief review of selected optics measurement methods is presented.

3.3.1 Non-TBT-based techniques

Quadrupole gradient modulation

The averaged beta function at a quadrupole can be measured by the quadrupole gradient modulation method. According to Eq. (2.42), tune shifts due to a perturbative localized integrated quadrupole gradient error $\Delta(K_1 L)$ is given by

$$\Delta\nu_y = \frac{1}{4\pi}\bar{\beta}_y\Delta(K_1 L),\tag{3.2}$$

where $\bar{\beta}_y$ is the average beta function at the quadrupole. According to Eq. (3.2), $\bar{\beta}_y$ can be derived from the slope of a linear fit of a set of data points $(\Delta(K_1 L), \Delta\nu_y)$. Figure 3.13 shows a screen shot of an application using quadrupole gradient modulation method to measure the average beta functions at the quadrupoles Q1 of the triplet and the derived beta functions at the IP and the waist. The blue and red curves in Fig. 3.13 shows the evolutions of horizontal and vertical betatron tunes due to modulation of the Q1 quadrupoles, respectively. This method is only applicable

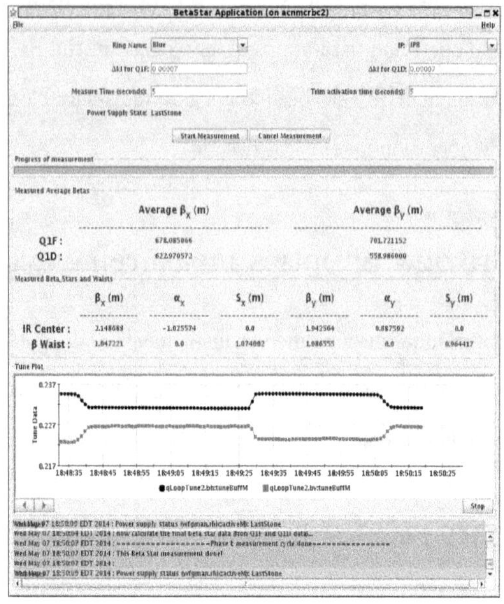

Figure 3.13: Screen shot of measurement results of average beta
function at triplet quadrupole Q1 and derived beta
waist using quadrupole gradient modulation method
(Courtesy of Y. Luo).

for quadrupoles with independent power supplies. It also requires the closed orbit is

not altered by the perturbation of quadruple integrated strength. Otherwise, part of

the measured tune shift could be caused by the closed-orbit variation at sextupoles

elsewhere in the accelerator. Assuming closed orbit is centered at the quadrupole to be

perturbed, the accuracy of the measured beta functions depends on the uncertainties of trim value $\Delta(K_1 L)$ and tune measurement.

Orbit response matrix (ORM) method

Equation (2.41) states that the closed orbit in a synchrotron is a superposition of dipole field errors propagated by the Green's function of Hill's equation. Consider a set of dipole perturbations given by $\theta_j, j = 1, 2, \cdots, N_b$, where N_b is the number of dipole kickers. The measured closed orbit y_i at the i-th BPM from the dipole perturbation is

$$y_i = \mathbf{R}_{ij}\theta_j, \quad i = 1, 2, \cdots, N_m, \quad j = 1, 2, \cdots, N_b, \tag{3.3}$$

where N_m is the number of BPMs which can be different from N_b, and the response matrix $\mathbf{R}$ is equal to Green's function G of Eq. (2.41) plus a contribution from the momentum shift associated with changing each dipole kicker [49].

An experimental ORM $\mathbf{R}_{\text{exp}}$ can be measured by perturbing known dipole kickers in a synchrotron and calculating the unit orbit response to an individual perturbation. A model ORM $\mathbf{R}_{\text{model}}$ can also be calculated from a designed lattice by an optics calculation engine like MADX and ELEGANT. The idea of the ORM method is to minimize the difference between $\mathbf{R}_{\text{exp}}$ and $\mathbf{R}_{\text{model}}$ by adjusting the parameters in the designed lattice such as the dipole and quadrupole field strengths, dipole as well as quadrupole rolls, calibration factors of dipole kickers and BPM gain factors. Among the many ORM-based codes, linear optics from closed orbit (LOCO) [50] has been widely applied to many synchrotron such as SPEAR3 [51], ALBA [52], SOLEIL [53] and the Fermilab Booster [54] for successful lattice modeling.

In fact, measuring $\mathbf{R}_{\text{exp}}$ can be very time consuming. Fitting $\mathbf{R}_{\text{model}}$ to $\mathbf{R}_{\text{exp}}$ for a complicated synchrotron accelerator can be a large-scale nonlinear optimization problem which requires very intense numerical computations and well behaved optimization algorithms.

3.3.2 TBT-based techniques

Properties of TBT BPM data

TBT BPM data measures transverse beam centroid displacement. For a stable beam, the centroid motion is a small amplitude oscillation around the closed orbit so that the data recorded by a BPM is usually buried in the BPM noise floor. To obtain meaningful TBT BPM data, a beam must be excited by a pulse kicker to perform decaying free betatron oscillation, or by an AC dipole to undergo driven betatron oscillation.

For a kick angle θ applied on a beam with a beta function $\beta_{y,0}$ at the kicker location, the excited initial amplitude of the decaying free betatron oscillation normalized to the local rms beam size $\sigma_{y,0}$ is $Z = \beta_{y,0}\theta/\sigma_{y,0}$. The decay of beam centroid motion for free betatron oscillation originates from the tune spread of the beam. Consider a Gaussian beam with rms beam width σ_y and rms momentum spread σ_δ. The tune shift of a single particle with a Courant-Snyder constant ϵ_y due to amplitude detuning and chromaticity is $\Delta\nu_y = -\nu_y'\epsilon_y/\epsilon_{y,\mathrm{rms}} + C_y\delta$, where ν_y' is the rms amplitude detuning and C_y is the chromaticity. Assuming the beam transverse and longitudinal distribution are uncorrelated, the TBT data recorded at the i-th BPM for the j-th turn has the form of [41]

$$y_i(j) = \theta\sqrt{\beta_{y,i}\beta_{y,0}}\,A_s(j)A(j)\cos[2\pi\nu_y j + \psi_{y,i} + \Delta\bar\phi(j)], \tag{3.4}$$

with

$$A_s(j) = \exp(-\frac{[\alpha(j)]^2}{2}), \quad \alpha(j) = \frac{2C_y\sigma_\delta \sin(\pi\nu_s j)}{\nu_s} \tag{3.5}$$

$$A(j) = \frac{1}{1 + (Q_p j)^2}\exp\left[-\frac{(Q_g j)^2}{2[1 + (Q_p j)^2]}\right] \tag{3.6}$$

$$\Delta\bar\phi(j) = -\frac{Z^2}{2}\frac{Q_p j}{1 + (Q_p j)^2} - 2\arctan(Q_p j) \tag{3.7}$$

where $\beta_{y,i}$ and $\psi_{y,i}$ are correspondingly the beta and phase function at the i-th BPM, ν_s is synchrotron tune, $Q_p = 4\pi\nu'_y$ and $Q_g = ZQ_p$ are the characteristic tunes. $\beta_{y,i}$ and $\psi_{y,i}$ can be derived from the amplitude and phase of the decaying simple harmonics oscillation. From Eqs. (3.4), $\beta_{y,i}$ and $\psi_{y,i}$ can be derived from the amplitude and phase of the decaying free betatron oscillation. The detuning parameters have to be well adjusted for a useful TBT BPM data of free betatron oscillation. This is because a better measurement resolution requires a longer decay time.

On the other hand, an AC dipole drives the driven betatron oscillation with a time-varying angular kick

$$\theta(t) = \theta_d \cos(2\pi\nu_d t + \phi_d), \tag{3.8}$$

where θ_d is the kick amplitude which is an adiabatic parameter, ν_d is the driving frequency of the AC dipole, and ϕ_d is the initial phase of the AC dipole. When the AC dipole driving tune ν_d is the same as the free betatron oscillation tune ν_f, there will be an on-resonance growth of the driven oscillation amplitude which causes beam loss. For a lepton machine such as a synchrotron light source, on-resonance driven oscillation is achievable due to the synchrotron radiation damping. However, for a hadron machine like RHIC, there is no such strong damping mechanism to counter balance the on-resonance amplitude growth so an AC dipole only drives off-resonance excitation. The TBT data for off-resonance driven betatron oscillation recorded at the i-th BPM for the j-th turn is given by (see Appendix A)

$$y_i(j) = A_d \sqrt{\beta_{d,i}} \cos(2\pi\nu_d j - \phi_d + \psi_{d,i}) \tag{3.9}$$

with

$$A_d = \frac{\theta_d \sqrt{\beta_{f,0}(1 - \lambda^2)}}{4\sin(\pi Q_-)}, \tag{3.10}$$

$$\beta_{d,i} = \frac{1 + \lambda^2 - 2\lambda\cos[2(\psi_{f,i} - \nu_f)]}{1 - \lambda^2}\beta_{f,i}, \tag{3.11}$$

$$\lambda = \frac{\sin(\pi Q_-)}{\sin(\pi Q_+)}, \quad Q_\pm = \nu_d \pm \nu_f, \tag{3.12}$$

$$\tan(\psi_{d,i} - \pi\nu_d) = \frac{1 + \lambda}{1 - \lambda}\tan(\psi_{f,i} - \pi\nu_f), \tag{3.13}$$

where $\beta_{f,0}$ is the beta function at the location of the AC dipole and ν_f is the tune of free betatron oscillation. In Eqs. (3.8) to (3.13), the subscript denoting transverse direction has been omitted for simplicity. The subscript "d" indicates quantities related to driven oscillation, while quantities with the subscript "f" pertain to free oscillation. Equations (3.8) to (3.13) state that the driven oscillation is a non-decaying betatron oscillation with modified beta function $\beta_{d,i}$ and phase function $\psi_{d,i}$. Therefore, direct measurement of amplitude and phase of driven betatron oscillation introduces systematic beta-beat and phase-beat. Additional considerations have to be taken into account to minimize these systematic measurement errors.

For convenience, TBT data from different BPMs are usually arranged into a matrix form. Consider M BPMs for a total of N turns data, the TBT BPM data matrix $\mathbf{Y}$ is defined as

$$\mathbf{Y} = \begin{pmatrix} y_1(1) & y_1(2) & \cdots & y_1(N) \\ y_2(1) & y_2(2) & \cdots & y_2(N) \\ \vdots & \vdots & \ddots & \vdots \\ y_M(1) & y_M(2) & \cdots & y_M(N) \end{pmatrix}, \tag{3.14}$$

where $y_i(j)$ is the output of the i-th BPM at the j-th turn. For a linear BPM system, the TBT BPM data matrix can be considered as a mixture of physical source signals which are usually harmonics oscillations with characteristic frequencies and BPM noise, i.e.,

$$\mathbf{Y} = \sum_i \mathbf{Y}_i(\nu_i) + \mathbf{N}, \tag{3.15}$$

where $\mathbf{Y}_i(\nu_i)$ is the i-th narrow-band source signal with the characteristic frequency ν_i and $\mathbf{N}$ is BPM noise. The frequency ν_i's are the betatron tunes, synchrotron tunes, higher harmonics of betatron tunes from nonlinear beam motion, etc. The BPM noise $\mathbf{N}$ is usually modeled as Gaussian white noise.

Numerical Analysis of Fundamental Frequency (NAFF)

NAFF is an algorithm for determining the frequency components of a quasi-periodic signal. It was originally developed for analysis of chaotic dynamical systems [55]. Later, it was introduced to the particle accelerator community to develop the frequency map analysis (FMA) for beam dynamics [56]. Figure 3.14 shows comparisons of relative errors in amplitude $\Delta A/A$ and frequence $\Delta f/f$ determination for a sinusoidal signal using NAFF and FFT. Since NAFF provides considerably more accurate estimation of frequency and amplitude compared to the conventional FFT, it can also be applied to TBT-based optics measurement [57]. However, the performance of NAFF is prone to BPM noise so pre-processing of TBT BPM data is usually required.

Model independent analysis (MIA)

MIA is a method for TBT BPM data analysis to untangle the eigenmodes by using statistical methods such as the singular value decomposition (SVD) [58]. The eigenmodes are identified as the uncorrelated eigenvectors which maximize the amount of variance of the TBT BPM data. Using the spatial and temporal functions from the eigenmodes, MIA is able to identify the physical source signals $\mathbf{Y}_i(\nu_i)$ in Eq. (3.15). The source signal of betatron motion provides information from which the linear optics can be derived. MIA has been applied to study beam dynamics in high energy

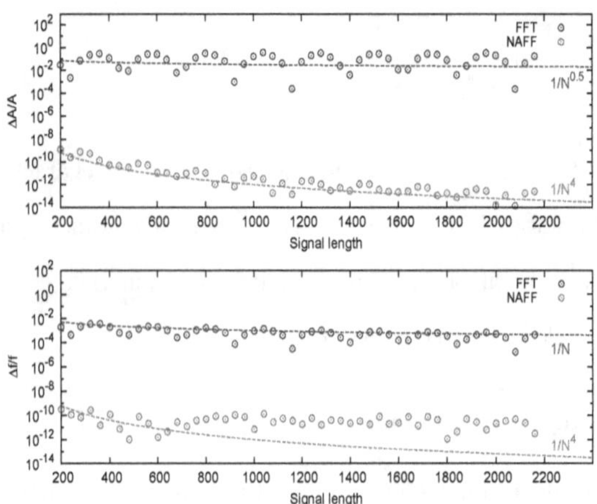

Figure 3.14: Comparison of amplitude A (top) and frequency f (bottom) accuracy of NAFF and FFT for a sinusoidal signal versus different signal lengths N.

synchrotrons, such as the Advanced Photon Source (APS) [59]. However, MIA is essentially a principal component analysis (PCA) technique which is prone to degenerate eigenmodes and modes with mixed source signals may be present in the MIA results such that the measurement accuracy may be degraded [27].

Independent component analysis (ICA)

ICA is another TBT BPM data processing technique which provides a remedy for MIA's limitation. ICA introduces the criterion of independence instead of uncorrelation so that the source signals separated by ICA are more immune to mode mixing and BPM noise. As an example, Fig. 3.15 shows the spatial and temporal functions of PCA and ICA analysis of simulated TBT BPM data. The simulated data is composed of simple harmonic oscillation with a tune of $\nu_1 = 8.705$ and random amplitudes at each of the 40 BPMs and a simple harmonic oscillation with a tune

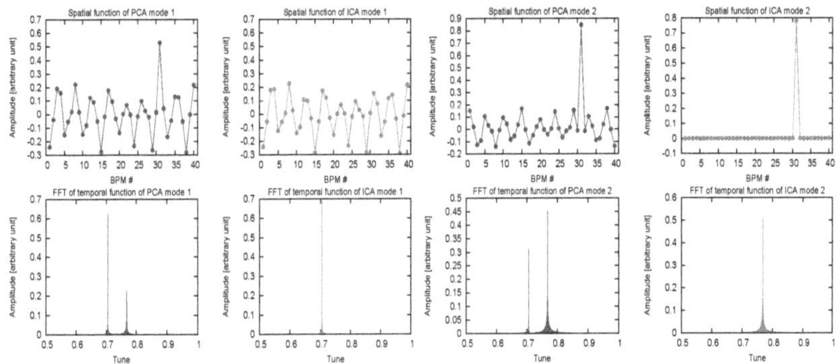

Figure 3.15: Spatial function (top) and FFT of temporal function (bottom) for modes of PCA (blue) and ICA (red).

of $\nu_2 = 1.768$ at the 31st BPM. The FFT spectra of the temporal functions of PCA shows mixed frequencies of ν_1 and ν_2, while in the ICA temporal functions ν_1 and ν_2 are well separated. ICA successfully identifies the simple harmonics with ν_2 at the 31st BPM in the spatial function of mode 2 and obtains the correct spatial function of simple harmonics with ν_1 in mode 1, while PCA cannot. The interpretation of spatial and temporal function will be discussed in detailed in Section 3.4.

Because of the outstanding performance of ICA in separation of independent source signals, there have been many applications of ICA to beam dynamics study, such as analysis of free betatron oscillation TBT BPM data for linear optics measurement at the Fermilab Booster [27] and digitized bunch slice signals analysis for study of beam dynamics within a long bunch in the Los Alamos Proton Storage Ring (PSR) [30]. There are also feasibility studies of applying ICA to free betatron oscillation TBT BPM data for measurement of linear coupling, dispersion function, and nonlinear magnet strengths [28, 29].

3.4   ICA for TBT BPM data analysis at RHIC

ICA is a data analysis technique to uncover the underlying random variables or source signals from multivariate samples. It considers the samples as linear mixtures of the random variables or source signals. A good example to illustrate the ICA data model is the cocktail-party problem. Consider N people speaking simultaneously in a party and their voices are recorded by M microphones at different locations. The signal sampled by each microphone is a linear mixture of the voices or sources depending on their distances from the microphone and background noise signals depending on the performance of the microphone. Denote the digitized record at the i-th microphone as $\mathbf{y}_i = \big(y_i(1), y_i(2), \cdots, y_i(T)\big)$ which is a row vector with T elements, noise signals at the i-th microphone as $\mathbf{n}_i = \big(n_i(1), n_i(2), \cdots, n_i(T)\big)$, and the j-th source signals as $\mathbf{s}_j = \big(s_j(1), s_j(2), \cdots, s_j(T)\big)$, then

$$\begin{pmatrix} \mathbf{y}_1 \\ \mathbf{y}_2 \\ \vdots \\ \mathbf{y}_M \end{pmatrix} = \begin{pmatrix} a_{11} & a_{12} & \cdots & a_{1N} \\ a_{21} & a_{22} & \cdots & a_{2N} \\ \vdots & \vdots & \ddots & \vdots \\ a_{M1} & a_{M2} & \cdots & a_{MN} \end{pmatrix} \begin{pmatrix} \mathbf{s}_1 \\ \mathbf{s}_2 \\ \vdots \\ \mathbf{s}_N \end{pmatrix} + \begin{pmatrix} \mathbf{n}_1 \\ \mathbf{n}_2 \\ \vdots \\ \mathbf{n}_M \end{pmatrix} \tag{3.16}$$

where the mixing coefficients a_{ij} are real numbers. Eq. (3.16) can be expressed in a matrix form as

$$\mathbf{Y} = \mathbf{AS} + \mathbf{N}, \tag{3.17}$$

where $\mathbf{Y} = (\mathbf{y}_1^T, \mathbf{y}_2^T, \cdots, \mathbf{y}_M^T)^T$ is the data matrix sampled at the microphones, $\mathbf{A}$ is the $M \times N$ mixing matrix containing the mixing coefficients with $\mathbf{A}_{ij} = a_{ij}$, $\mathbf{S} = (\mathbf{s}_1^T, \mathbf{s}_2^T, \cdots, \mathbf{s}_N^T)^T$ is the source matrix, and $\mathbf{N} = (\mathbf{n}_1^T, \mathbf{n}_2^T, \cdots, \mathbf{n}_M^T)^T$ is the noise matrix. The superscript "T" denotes the matrix transpose. The case with $M \geq N$ is to be explored in this study.

The mission of ICA is to separate the source matrix $\mathbf{S}$ and mixing matrix $\mathbf{A}$ directly from the data $\mathbf{Y}$ without a priori knowledge of detailed conditions such as where the people or the microphones are. Therefore, ICA belongs to the blind source separation (BSS) problem. In general, $\mathbf{Y}$ and $\mathbf{S}$ can be random variables but not necessarily signals.

The only assumption of ICA for separation of source signals is that the source signals $\mathbf{s}_j$'s are mutually independent. Statistically, n random variables $X_1, X_2, \cdots, X_n$ are mutually independent if and only if the joint probability density function (pdf) $p_{X_1, X_2, \cdots, X_n}(x_1, x_2, \cdots, x_n)$ is related to the marginal pdf of X_i's denoted as $p_{X_i}(x_i)$ as

$$p_{X_1, X_2, \cdots, X_n}(x_1, x_2, \cdots, x_n) = p_{X_1}(x_1)p_{X_2}(x_2) \cdots p_{X_n}(x_n). \tag{3.18}$$

In practice, the statistical expectation value $E[X] = \int_{-\infty}^{\infty} x p_X(x) dx$ is a measurable to quantify the independent relation. For example, two random variables X_1 and X_2 are statistically independent if and only if

$$\mathrm{Cov}[f_1(X_1), f_2(X_2)] = E[f_1(X_1)f_2(X_2)] - E[f_1(X_1)]E[f_2(X_2)] = 0, \tag{3.19}$$

where f_1, f_2 are arbitrary non-diverging functions and $\mathrm{Cov}[f_1(X_1), f_2(X_2)]$ is the covariance of $f_1(X_1)$ and $f_2(X_2)$. If Eq. (3.19) holds for the special case when $f_1(X_1) = X_1$ and $f_2(X_2) = X_2$, then X_1 and X_2 are uncorrelated.

If the source signals have time correlation and non-overlapping as well as narrow-band power spectra, like the voice signals in the cocktail party problem, an alternative quantity to measure independence is the time-lagged covariance defined for two signals $\mathbf{y}_1(t)$ and $\mathbf{y}_2(t)$ as

$$C_\tau(\mathbf{y}_1, \mathbf{y}_2) \equiv \langle \mathbf{y}_1(t)\mathbf{y}_2(t+\tau)\rangle, \tag{3.20}$$

where τ is called the time-lagged parameter, $\langle \cdots \rangle$ denotes the expectation value over time, and it has been assumed that $\langle \mathbf{y}_1(t)\rangle = \langle \mathbf{y}_2(t)\rangle = 0$. The time-lagged covariance of a single signal, $C_\tau(\mathbf{y}) = \langle \mathbf{y}(t)\mathbf{y}(t+\tau)\rangle$, is called auto-covariance. The auto-covariance of a Gaussian white noise signal $\mathbf{n}(t)$ is null, i.e., $C_\tau(\mathbf{n}) = 0$ holds for every τ. Source signals $\mathbf{s}_1$ and $\mathbf{s}_2$ are independent if

$$C_\tau(\mathbf{s}_1, \mathbf{s}_2) = \langle \mathbf{s}_1(t)\mathbf{s}_2(t+\tau) \rangle = 0 \tag{3.21}$$

holds for every τ.

The criterion given by Eqs. (3.19) and (3.21) provides different cost functions to measure statistical independence. There are many types of ICA algorithms depending on the cost function adopted and method used to optimize the cost function. For example, the joint approximate diagonalization of eigen-matrices (JADE) algorithm separates the source signals by jointly diagonalize the fourth cumulant matrices [60], the FastICA algorithm by choosing robust nonlinear functions as the contrast and using a fixed point algorithm to optimize the cost function [61] and the second order blind identification (SOBI) by joint diagonalization of a set of time-lagged covariance matrices [62]. Since the source signals of the TBT BPM data are narrow-band beam oscillations with characteristic frequencies as described by Eq. (3.15), the SOBI ICA algorithm is particularly suitable for the TBT BPM data analysis. In the next section, the principles of SOBI ICA algorithm will be presented.

3.4.1 Principle of SOBI algorithm

The SOBI ICA algorithm starts with the linear mixture model from Eq. (3.17) where the source signals are assumed to be mutually independent and temporally correlated. Consequently, the time-lagged covariance matrix $\mathbf{C}_\tau(\mathbf{S}) \equiv \langle \mathbf{S}(t)\mathbf{S}(t+\tau)^T \rangle$ is diagonal, i.e., $C_\tau(\mathbf{s}_i, \mathbf{s}_j) = S_i(\tau)\delta_{ij}$, where τ is the time-lagged constant and δ_{ij} is the Kronecker delta function. At the same time, only the zero time-lagged auto-covariance of the noise signals survive in the time-lagged covariance matrices $\mathbf{C}_\tau(\mathbf{N})$ for the noise data $\mathbf{N}$, the time-lagged covariance matrices $\mathbf{C}_\tau(\mathbf{Y})$ for data matrix $\mathbf{Y}$ in Eq. (3.17) is

$$\mathbf{C}_\tau(\mathbf{Y}) = \mathbf{A}\mathbf{C}_\tau(\mathbf{S})\mathbf{A}^T + \mathrm{diag}(\sigma_1^2, \sigma_2^2, \cdots, \sigma_M^2)\delta_{\tau 0}, \tag{3.22}$$

where $\mathrm{diag}(\cdots)$ denotes a diagonal matrix with numbers in the parenthesis being the diagonal elements and $\sigma_i = C_{\tau=0}(\mathbf{n}_i)$ is the zero time-lagged auto covariance, or the standard deviation, of the Gaussian white noise at the i-th BPM. Since all of the time-lagged covariance matrices $\mathbf{C}_\tau(\mathbf{S})$'s are supposed to be diagonal, the mixing matrix $\mathbf{A}$ is found as the joint diagonalizer of $\mathbf{C}_\tau(\mathbf{Y})$. The source matrix $\mathbf{S}$ can then by obtained from the data matrix $\mathbf{Y}$ using a de-mixing matrix which is calculated as an inverse linear transformation of $\mathbf{A}$. The detailed algorithm of SOBI ICA is discussed as follows.

First, a whitening procedure is applied to preprocess the raw data matrix $\mathbf{Y}$. The zero time-lagged covariance matrix $\mathbf{C}_{\tau=0}(\mathbf{Y}) = \langle \mathbf{Y}(t)\mathbf{Y}(t)^T \rangle$ is decomposed by the singular valued decomposition (SVD) as

$$\mathbf{C}_{\tau=0}(\mathbf{Y}) = \mathbf{U}\mathbf{\Lambda}\mathbf{U}^T, \tag{3.23}$$

where $\mathbf{U}$ is an $M \times M$ othogonal matrix whiose columns are the eigenvectors of $\mathbf{C}_{\tau=0}(\mathbf{Y})$, and $\mathbf{\Lambda} = \mathrm{diag}(\lambda_1, \lambda_2, \cdots, \lambda_M)$ is an $M \times M$ diagonal matrix whose diagonal elements are the corresponding eigenvalues. The eigenvalues in $\mathbf{\Lambda}$ are also called singular values and they are arranged in a non-ascending sequence, i.e., $\lambda_1 \geq \lambda_2 \geq \cdots \geq \lambda_M$. The amplitude of a singular value indicates the variance of $\mathbf{C}_{\tau=0}(\mathbf{Y})$ projected onto the axis defined by the corresponding eigenvector. From the information point of view, the decomposition of $\mathbf{C}_{\tau=0}(\mathbf{Y})$ on the axis of an eigenvector with a larger singular value contains more important information, while the information projected onto the axes of an eigenvector with small singular values are redundant. Equation (3.23) can be rewritten as

$$\mathbf{C}_{\tau=0}(\mathbf{Y}) = (\mathbf{U}_1, \mathbf{U}_2) \begin{pmatrix} \mathbf{\Lambda}_1 & 0 \\ 0 & \mathbf{\Lambda}_2 \end{pmatrix} \begin{pmatrix} \mathbf{U}_1^T \\ \mathbf{U}_2^T \end{pmatrix}, \tag{3.24}$$

where $\mathbf{\Lambda}_1$ is a diagonal matrix containing the first n_c largest number of singular values in $\mathbf{\Lambda}$, and $\mathbf{\Lambda}_2$ is a diagonal matrix containing the other singular values. n_c

is also called cut-off mode number. $\mathbf{U}_1$ is an $M \times n_c$ orthogonal matrix containing the first n_c columns of $\mathbf{U}$, and $\mathbf{U}_2$ is an orthogonal matrix containing the remaining columns of $\mathbf{U}$. A whitening matrix can be constructed as

$$\mathbf{V} = \Lambda_1^{-1/2}\mathbf{U}_1^T \tag{3.25}$$

such that the whitened data $\mathbf{Z} = \mathbf{V}\mathbf{X}$ has a zero time-lagged covariance $C_{\tau=0}(\mathbf{Z}) = \mathbf{I}$, where $\mathbf{I}$ is the $l \times l$ identity matrix. Therefore $\mathbf{Z}$ is spatially white. The whitening procedure is also a PCA algorithm. It removes redundant information and noise from the raw data, de-correlates and normalizes the data to facilitate the next step.

Using the whitened data $\mathbf{Z}$ and a set of time-lag constants $\{\tau_k\}$ $(k = 1, 2, \ldots, K)$, the time-lagged covariance matrices are computed as $\{\mathbf{C}_{\tau_k}(\mathbf{Z}) = \langle \mathbf{Z}(t)\mathbf{Z}(t + \tau_k)^T \rangle\}$. The symmetrized time-lagged covariance matrices are formed as $\overline{\mathbf{C}}_{\tau_k}(\mathbf{Z}) = [\mathbf{C}_{\tau_k}(\mathbf{Z}) + \mathbf{C}_{\tau_k}(\mathbf{Z})^T]/2$ such that they are real and symmetric and thus their eigenvalue decompositions are well defined. At last, a Jacobi-like algorithm [63] is $\overline{\mathbf{C}}_{\tau_k}(\mathbf{Z})$ such that

$$\overline{\mathbf{C}}_{\tau_k}(\mathbf{Z}) = \mathbf{W}\mathbf{D}_k\mathbf{W}^T, \tag{3.26}$$

where $\mathbf{D}_k$'s are diagonal matrices. Finally, the estimated source signals are given by $\mathbf{S} = \mathbf{W}^T\mathbf{V}\mathbf{Y}$ and the mixing matrix by $\mathbf{A} = \mathbf{V}^{-1}\mathbf{W}$. By carefully choosing the number n_c to preserve relevant singular values, all source signals can be correctly collected into $\mathbf{S}$ such that

$$\mathbf{Y} = \mathbf{A}\mathbf{S} + \mathbf{N}, \tag{3.27}$$

where $\mathbf{N} = \mathbf{Y} - \mathbf{A}\mathbf{S}$ is the estimated noise signals. Since a pairs of mutually reciprocal scaling factors $(\kappa, 1/\kappa)$ can be applied to $\mathbf{A}$ and $\mathbf{S}$ such that the product of $\mathbf{A}$ and $\mathbf{S}$ is kept unchanged, the mixing matrix and source signals are not unique. Therefore, all the $\mathbf{A}$'s and $\mathbf{S}$'s are said to be equivalent up to a non-zero scaling factor.

For demonstration, a noise-free cocktail-party problem with 2 source signals and 2 microphones is studied. The two source signals are manually mixed to generate two mixture signals to simulate the recorded signals at two different microphones. The mxiture signals are then analyzed by ICA to separate the source signals. Figure 3.16 shows the digitized wave form of the source signals, mixture signals, the intermediate whitened signals and the separated source signals. The wave forms of the separated source signals are equivalent to those of the original source signals. A joint density

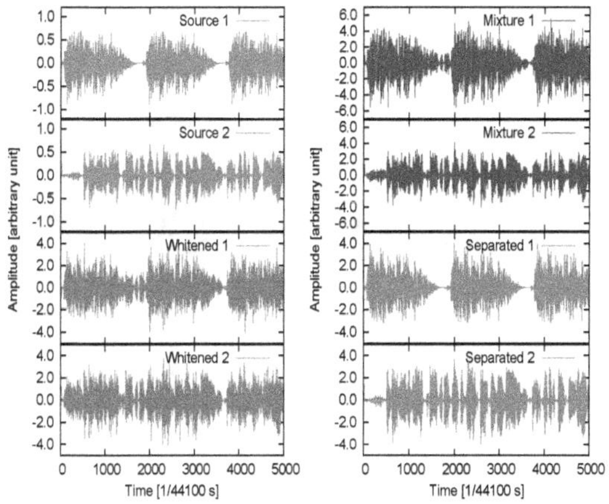

Figure 3.16: Digitized wave form of the source signals, mixture signals, whitened signals and separated source signals.

plot which draws points $\big(x = \mathbf{a}(t), y = \mathbf{b}(t)\big)$ as a function of t can show the relation between the two temporal signals $\mathbf{a}(t)$ and $\mathbf{b}(t)$. Figure 3.17 shows the joint density plots of the corresponding signals in Fig. 3.16. The joint density of the source signals shows a distribution with the shape of an upright cross, which indicates that a knowledge of one source signal tells no knowledge about the other. As shown by the joint density plot of the mixture signals, the effect of mixing is to scale, rotate and tilt the joint density distribution. The whitening procedure recovers all the features of the joint density distribution of the source signals except for a rotation. Finally, the

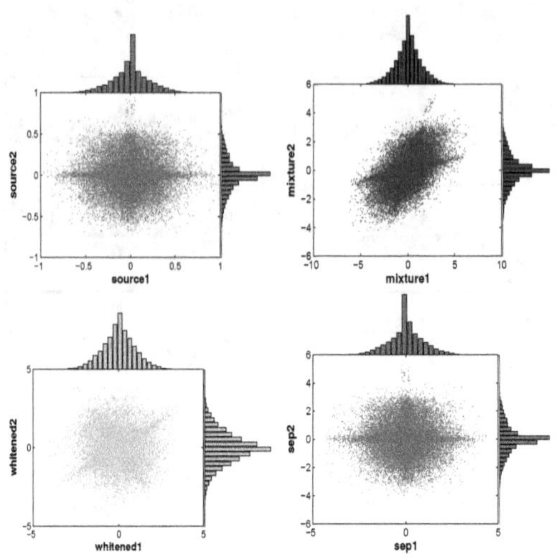

Figure 3.17: Joint density plots of corresponding signals in Fig. 3.16.

Jacobi-like algorithm achieves this rotation and correctly separates the source signals, as shown by the joint density of the separated source signals.

TBT BPM data is normally a linear mixture of signals originating from physical sources with their own characteristic frequencies, such as betatron oscillation, synchrotron oscillation, nonlinear beam motions and electronic noises. A physical separation of the BPM data into these physical sources is of great interest for deeper understanding of beam dynamics. The source signals separated by ICA have non-overlapping power spectra because of their properties of mutual independence. Therefore ICA is particularly suitable for the task of physical source separation of BPM data. Application of ICA to TBT BPM data $\mathbf{Y}$ extracts the mixing matrix $\mathbf{A}$ and $\mathbf{S}$. The i-th row of $\mathbf{S}$, which is denoted as $\mathbf{s}_i$, gives the i-th source signal or temporal function. The i-th column of $\mathbf{A}$, which is denoted as $\mathbf{A}_i$, is called the i-th spatial function. The spatial and temporal function pair $(\mathbf{A}_i, \mathbf{s}_i)$ is called the i-th mode. The correspondence of the i-th mode to a certain beam motion is identified by comparing the frequency of the temporal function $\mathbf{s}_i$ to the characteristic frequency of the beam motion, such as betatron tunes and synchrotron tune. Finally, the TBT BPM data $\mathbf{Y}$ can be decomposed into different modes as

$$\mathbf{Y} = \sum_{i=1}^{n_c} \mathbf{A}_i \mathbf{s}_i + \mathbf{N}, \tag{3.28}$$

where the noise signal $\mathbf{N}$ is very useful for estimation of BPM noise performance, and the spatial function $\mathbf{A}_i$ provides important information of beam motion corresponding to $\mathbf{s}_i$, such as beta and phase functions of betatron motion. The details of applications of SOBI for BPM noise estimation and beam optics measurement are separately addressed in the next two subsections.

3.4.2   Application of SOBI for BPM noise estimation

Many beam diagnostics techniques rely on the usage of BPMs with required measurement accuracy. Therefore, it is very important to have a good understanding of the limitations of BPM performance for a correct interpretation of the diagnostics results. At RHIC, there have been many efforts to study the BPM performance, such as the statistical identification of malfunctioning BPMs using the SVD technique [64]. However, a detailed study of BPM noise performance is highly desired for reliable TBT-based beam dynamics study. Since the SOBI ICA algorithm is particularly efficient and robust in extracting physical source signals from TBT BPM data, the resulting noise signal $\mathbf{N}$ in Eq. (3.28) gives a very good estimation of BPM noise and the standard deviation σ_i for the i-th estimated noise signal $\mathbf{n}_i$ quantifies the noise level at the i-th BPM.

Computer simulation was conducted to evaluate the performance of the SOBI ICA algorithm for BPM noise estimation. In the simulation, TBT BPM data was generated by single particle tracking using the SimTrack [65] through a lattice for 2013 RHIC polarized proton operation for 1024 turns at 159 BPMs. The lattice starts at the injection point IP6 and includes all elements along the ring in a clockwise sequence. The simulated beam oscillation amplitude was chosen to be about 300μm in the middle of the arc, which is a level of excited beam oscillation amplitude similar to that observed during experiments. For each BPM, Gaussian white noise which is generated from a unique random seed and a standard deviation σ_{set} was added to the simulated beam oscillation data. For each BPM, σ_{set} was sampled from a uniform distribution within the range $[0, 120\mu\mathrm{m}]$ which covers the experimentally estimated average RHIC BPM noise of 60μm [66]. The SOBI ICA algorithm was applied to the simulated TBT BPM data to obtain an estimation $\sigma_{\mathrm{estimated}}$ of the standard deviation of the simulated Gaussian white noise at each BPM.

The left plot of Fig. 3.18 shows the amplitude of singular values for all modes. The singular values for the first two modes dominate the others by at least 3 orders of magnitude. The FFT spectra of the temporal functions for the first 4 modes are shown in the right plot of Fig. 3.18, in which the temporal functions of the first two modes exhibit a clean frequency peak at the betatron oscillation tune, while the spectra of the other two temporal functions are pretty noisy. Therefore, the cut-off mode number is chosen to be $n_c = 2$, as shown in the left plot of Fig. 3.18.

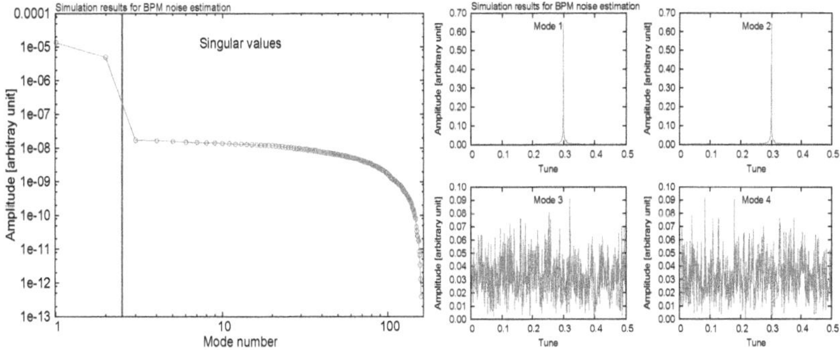

Figure 3.18: Singular values (left) and FFT spectra of the temporal functions of first 4 modes (right) for simulation.

The results for simulation of BPM noise estimation are shown in Fig. 3.19, in which the left plot shows the $\sigma_{estimated}$ and σ_{set} at different BPMs and the right plot shows the relation for $\sigma_{estimated}$ versus σ_{set}. $\sigma_{estimated}$ and its error bar are the average value and standard deviation of 10 estimations of BPM noise, respectively. The simulation results show that σ_{set} is successfully captured by $\sigma_{estimated}$ for all BPMs with a strong linear correlation correlation $r = 0.99982$.

Since its performance for BPM noise estimation had been proven to be outstanding by computer simulation, the SOBI ICA algorithm was applied to the TBT BPM data of AC dipole driven betatron oscillation taken during beam experiments in the RHIC

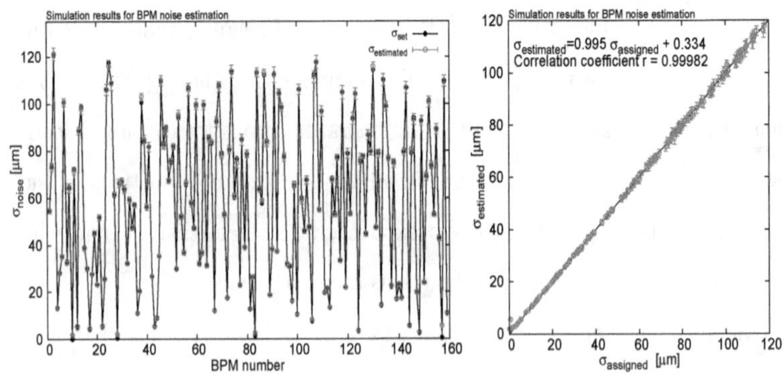

Figure 3.19: Singular values (left) and FFT spectra of the temporal functions of first 4 modes (right) for simulation.

2013 polarized proton operation for BPM noise estimation. An example of amplitudes
of singular values for a TBT BPM data is shown in the right plot of Fig. 3.20, in which
the singular values of the first two modes are dominating. The right plot shows the
FFT spectra of the first 4 modes. The spectra of the temporal functions of two first 2
modes peaks at the AC dipole driven frequency. The spectra of the temporal functions
of the 3rd and the 4th modes show characteristic frequencies of the synchrotron tune
and the second harmonics of synchrotron tune, respectively. The temporal functions
of the other modes are mostly dominated by broad-band noise.

Figure 3.21 shows the estimated BPM noise for the Blue ring for a beam optics
correction experiment of RHIC 2013 polarized proton operation with $n_c = 4$ and
$n_c = 6$ for comparison. The BPM noise distribution is consistent with the fact that
there are 4 types of BPMs as listed in Table 3.5. Extending n_c to include 6 modes
results in underestimation of noise at the triplets, which is not consistent with the
behavior of a physical source of beam motion. Therefore, the cut-off mode number
was chosen to be $n_c = 4$. Figure 3.22 shows the histograms of estimated BPM noise

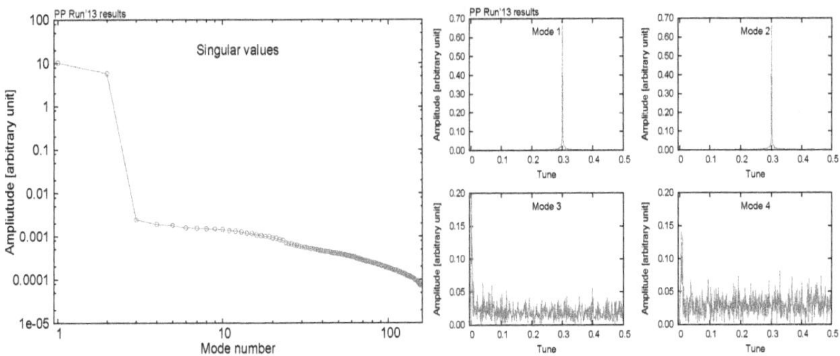

Figure 3.20: Typical singular values (left) and FFT spectra of the
temporal functions of first 4 modes (right) for TBT
BPM data of the 2013 polarized proton operation.

for the Blue ring corresponding to the case of $n_c = 4$ in Fig. 3.21. For both cases
of before and after optics correction, BPM noise shows $\sigma_{\text{Type 1}} < \sigma_{\text{Type 4}} < \sigma_{\text{Type 2}} <$
$\sigma_{\text{Type 3}} < 60$ μm. The average noise is larger at the time after optics correction, which
may come from the fact that the larger beam emittances and lower beam intensities
after optics correction, as listed in Table 3.6.

Table 3.6: Experimental parameters for PP Run'13.

		(ϵ_x, ϵ_z) [μm · rad]	Intensity [$\times 10^{11}$]	Bunches
Blue	Before correction	$(19.42, 15.61)$	1.39	6
	After correction	$(31.66, 19.22)$	1.38	6
Yellow	Before correction	$(12.75, 9.45)$	1.50	7
	After correction	$(22.25, 24.25)$	1.318	7

Figure 3.23 shows the estimated BPM noise for the Yellow ring in the same beam

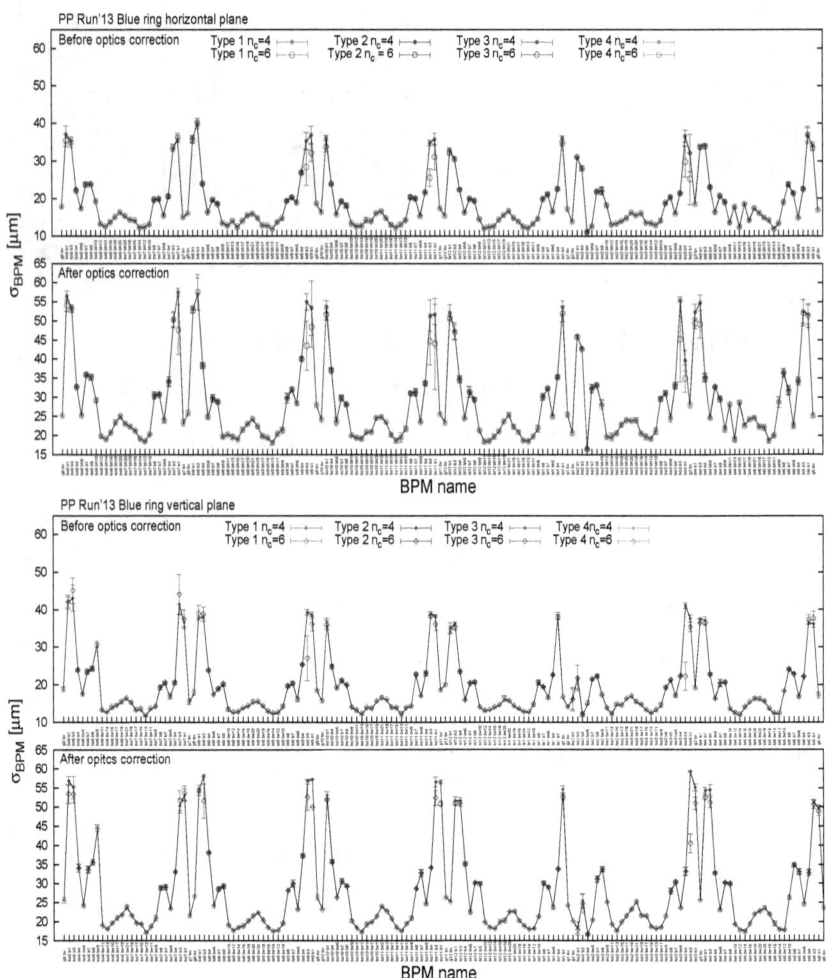

Figure 3.21: Estimated BPM noise for the Blue ring for the 2013 polarized proton operation with cut-off mode number $n_c = 4$ (solid marker) and $n_c = 6$ (hollow marker).

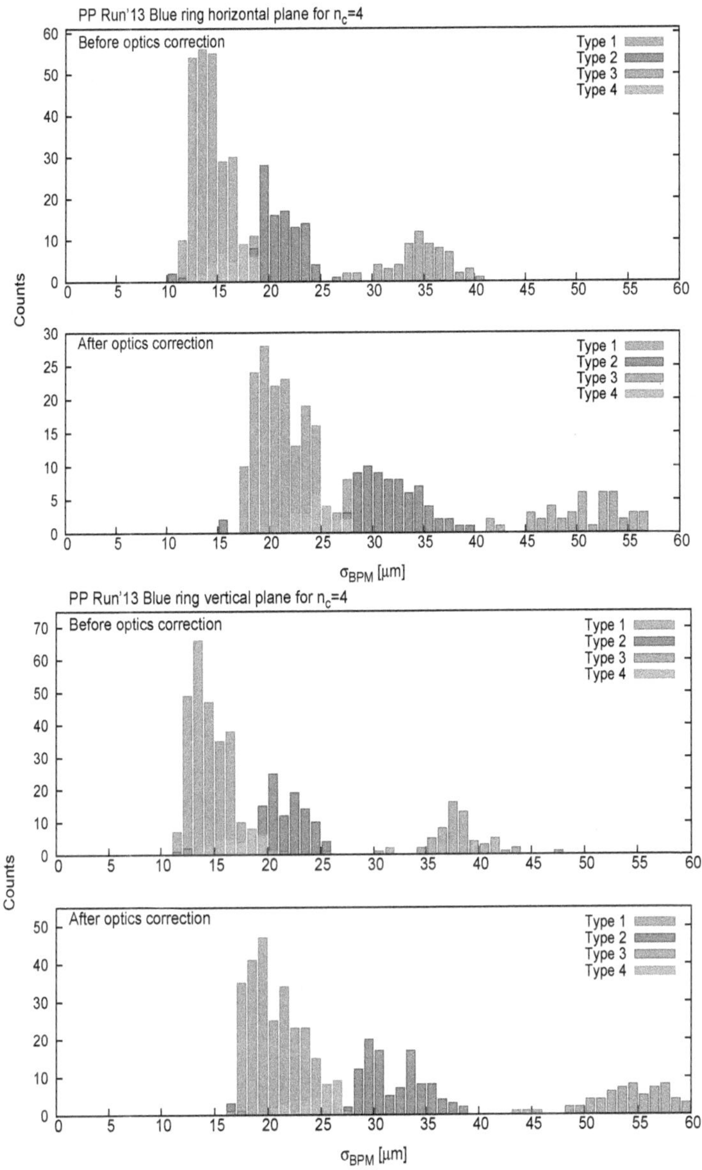

Figure 3.22: Histogram of the estimated BPM noise for the Blue ring for the 2013 polarized proton operation with $n_c = 4$.

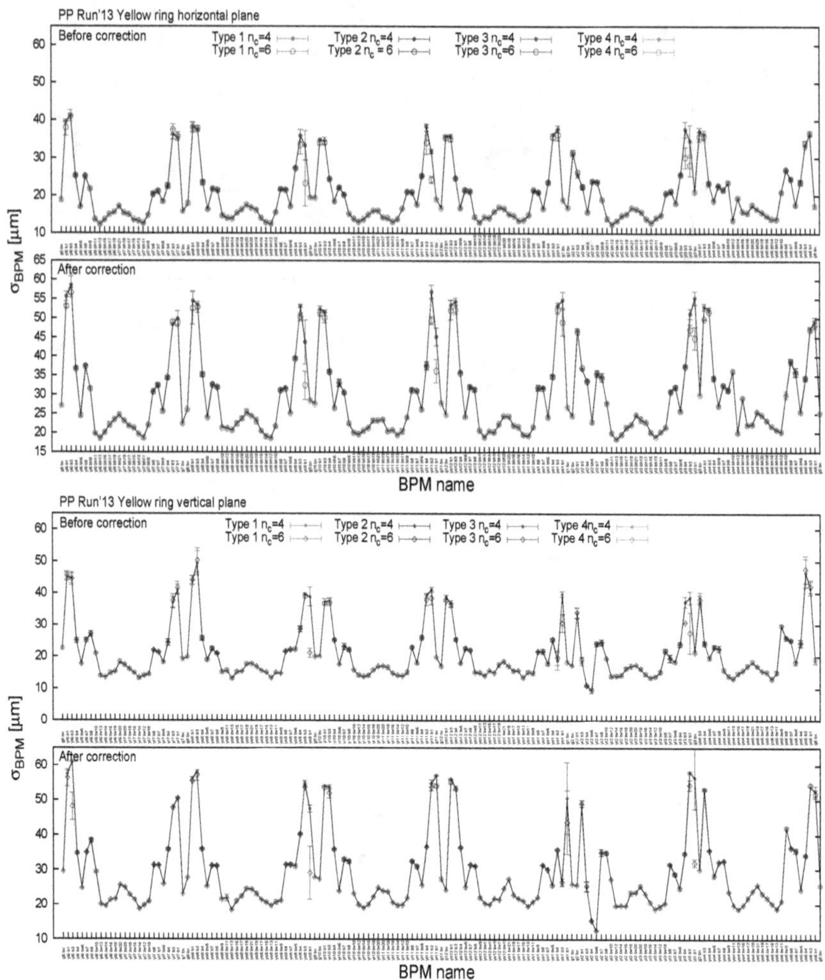

Figure 3.23: Estimated BPM noise for the Yellow ring for the 2013

polarized proton operation with cut-off mode number

$n_c = 4$ (solid marker) and $n_c = 6$ (hollow marker).

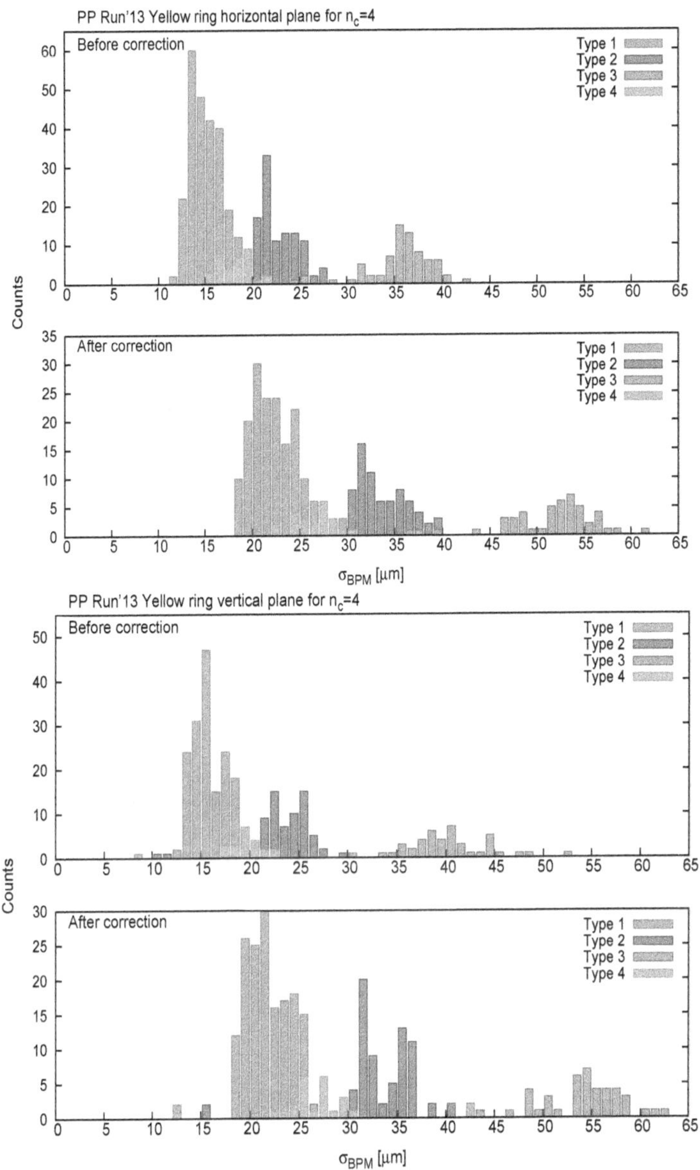

Figure 3.24: Histogram of the estimated BPM noise for the Yellow ring for the 2013 polarized proton operation with $n_c = 4$.

optics correction experiment with $n_c = 4$ and $n_c = 6$ for comparison. Again, the 5th and 6th modes do not represent physical sources of beam motion so that $n_c = 4$ was chosen. Figure 3.24 shows the corresponding histogram for Fig. 3.23, in which $\sigma_{\text{Type 1}} < \sigma_{\text{Type 4}} < \sigma_{\text{Type 2}} < \sigma_{\text{Type 3}} < 65$ μm. Similarly, the average noise is larger at the time after the optics correction. The reason may arise from the larger beam emittances and lower beam intensities after the optics correction as listed in Table 3.6.

Detailed understanding of the reasons for the observed BPM noise distribution requires further investigation such as how the BPM noise is related to the mechanical as well as electro-magnetic details of each BPM and noise figure of the electronics.

3.4.3 Application of SOBI ICA algorithm for optics measurement

The quality of linear beam optics parameters such as the beta and phase functions is of great importance for beam stability, luminosity and polarization performance of RHIC. Therefore, a prompt and accurate technique for optics measurement is highly desired to provide valuable information for optimization of machine performance. Since TBT BPM data contains fruitful information of beam dynamics, there have been many efforts to explore various methods for TBT BPM data based optics measurement, such as nonlinear fitting approach [67] and interpolated FFT technique [68,69]. Because of the outstanding performance, the SOBI ICA algorithm was introduced to RHIC for efficient and robust model-independent optics measurements in 2013.

According to Eq. (3.28), TBT BPM data are composed of different modes corresponding to beam oscillation at various characteristic frequencies. For each mode, its temporal function is related to a characteristic frequency of a beam oscillation, while its spatial function contains information about beam optics relevant to the corresponding beam oscillation. There are two modes for the betatron oscillation component of the TBT BPM data, i.e.,

$$\mathbf{Y}_{\text{betatron}} = \mathbf{A}_1\mathbf{s}_1 + \mathbf{A}_2\mathbf{s}_2, \tag{3.29}$$

where $\mathbf{s}_1, \mathbf{s}_2$ are the cosine-like and sine-like temporal functions whose FFT spectrum are peaked at the betatron tune, and $\mathbf{A}_1, \mathbf{A}_2$ are the betatron spatial functions related to the beta and phase functions as

$$\beta_i = \mathcal{F}(A_{1,i}^2 + A_{2,i}^2), \tag{3.30}$$

$$\psi_i = \arctan(\frac{A_{2,i}}{A_{1,i}}) + \mathcal{C}, \tag{3.31}$$

where β_i, ψ_i are the beta and phase function at the i-th BPM, respectively, $A_{1,i}, A_{2,i}$ are the i-th element of the spatial functions $\mathbf{A}_1, \mathbf{A}_2$, correspondingly, and $\mathcal{F}, \mathcal{C}$ are constants. The phase constant $\mathcal{C}$ will be cancelled when calculating the phase advance $\Delta\psi_{ij} = \psi_j - \psi_i$ between the i-th and j-th BPMs. The scaling factor $\mathcal{F}$ originates from the fact that the mixing matrix determined by ICA is not unique. It needs to be calibrated to obtain the absolute beta functions. This calibration can be realized by using the relationship of the beta and phase functions at two consecutive BPMs separated by a drift space, as depicted by Fig 3.25 in which β_1, β_2 are the absolute beta functions, $\tilde{\beta}_1, \tilde{\beta}_2$ are the uncalibrated beta functions, L and $\Delta\psi_{12}$ are the distance and the phase advance between the two BPMs, respectively. Using the relationship $\sqrt{\beta_1\beta_2}\sin\Delta\psi_{12} = L$, the scaling factor is obtained as

$$\mathcal{F} = \frac{L}{\sqrt{\tilde{\beta}_1\tilde{\beta}_2}\sin\Delta\psi_{12}}. \tag{3.32}$$

At each insertion region of RHIC, there are two pairs of BPMs separated by a drift space available for beta function calibration.

If analyzed TBT BPM data is for free betatron oscillation, Eqs. (3.30) and (3.31) give the unbiased beta function β_f and phase function ψ_f of the real machine, respectively. Because of the advantages of AC dipole driven betatron oscillation, TBT

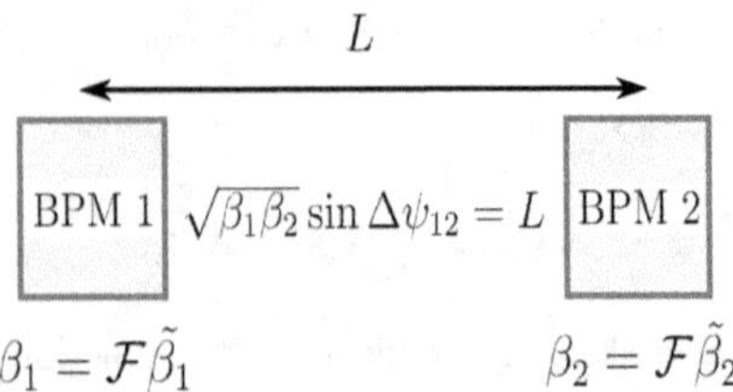

Figure 3.25: Beta function calibration using two BPMs separated
by a drift space.

BPM data of this type of excited beam oscillation is preferred for accurate optics measurement. However, when it comes to analysis of TBT BPM data for AC dipole driven betatron oscillation, the measured optics are the modified beta function β_d and phase function ψ_d in Eqs (3.11) and (3.13), respectively. β_d and ψ_d include systematic beta-beat and phase-beat errors depending on the distance between the AC dipole driven tune and the betatron tune $\Delta\nu = \nu_d - \nu_f$. For example, the red and blue curves in Fig. 3.26 show the systematic beta-beat in β_d comparing to β_f for $\Delta\nu_1 = 0.01$ and $\Delta\nu_2 = -0.01$, respectively. Denote $\psi_{ij} = \psi_j - \psi_i$ as the phase advance between the i-th and j-th BPMs, the relative phase-beat between the i-th and j-th BPM can be defined as $\Delta\psi_{ij} = \psi_{ij,\text{measured}} - \psi_{ij,\text{model}}$. For the lattice of RHIC 2013 polarized proton operation with $|\Delta\nu| = 0.01$, the maximum systematic error on beta-beat and relative phase-beat are approximately 7.0% and 5.5×10^{-3} rad, respectively. The systematic error in β_d and ψ_d can be minimized by averaging the measurement results with opposite tune distances, i.e., $\beta = 0.5 \times \left[\beta_d(\Delta\nu) + \beta_d(-\Delta\nu)\right]$, as shown by the green curve in Fig. 3.26, and $\psi_{ij} = 0.5 \times \left[\psi_{ij}(\Delta\nu) + \psi_{ij}(-\Delta\nu)\right]$. The maximum systematic error of beta-beat and relative phase-beat can be reduced to be about 0.3% and 1.0×10^{-3} rad, respectively.

To evaluate the performance of the averaging method for optics measurement,

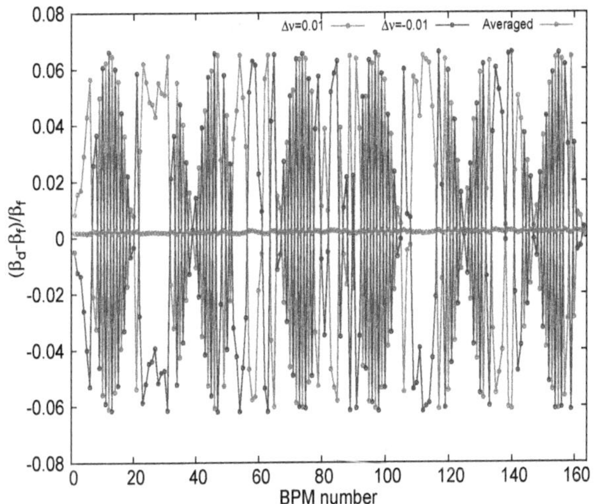

Figure 3.26: $(\beta_d - \beta_f)/\beta_f$ for $\Delta\nu = \pm 0.01$ and their average.

computer simulation was conducted to study the robustness of the proposed method against BPM random noise and calibration errors. Define the rms measurement errors σ_β/β and $\sigma_{\Delta\psi}$ as

$$\sigma_\beta/\beta \equiv \sqrt{\frac{1}{M}\sum_{i=1}^{M}\left(\frac{\beta_{\text{measured}} - \beta_{\text{model}}}{\beta_{\text{model}}}\right)^2}, \tag{3.33}$$

$$\sigma_{\Delta\psi} \equiv \sqrt{\frac{1}{M-1}\sum_{i=1}^{M-1}\left(\psi_{ij,\text{measured}} - \psi_{ij,\text{model}}\right)^2}, \tag{3.34}$$

where M is the total number of BPMs. The rms measurement errors σ_β/β and $\sigma_{\Delta\psi}$ are used to quantify the performance of the averaging method. In the simulation, a single particle was tracked using SimTrack [65] in a lattice for RHIC 2013 polarized proton operation by 1024 turns under the influence of adiabatic AC dipole excitation to generate TBT BPM data. Either white Gaussian BPM noise with standard deviation σ_{noise} or Gaussian random BPM calibration error with standard deviation σ_{cal} is assigned to the TBT data at each BPM. Figure 3.27 shows the rms measurement errors for different BPM noise levels with σ_{noise} ranging from 0 to 100 μm. The residual

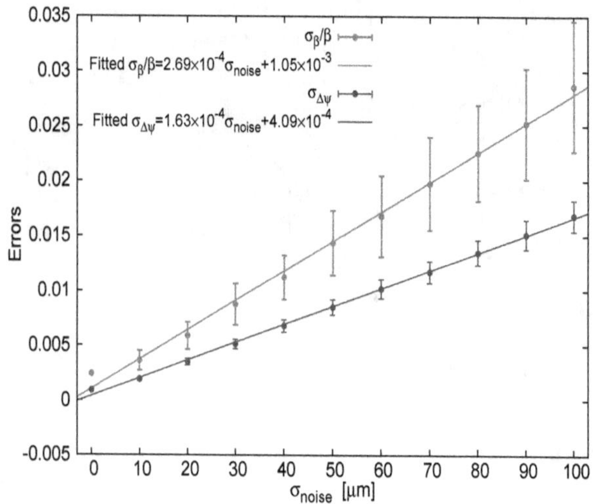

Figure 3.27: Estimation of rms errors σ_β/β and $\sigma_{\Delta\psi}$ with various BPM random noise levels σ_{noise}.

rms measurement errors at $\sigma_{\text{noise}} = 0$ are negligible compared to those for a realistic noise level of 60 μm found by the BPM noise estimation. The ratio of the slope of the linear fitting for σ_β/β to that for $\sigma_{\Delta\psi}$ is $2.69 \times 10^{-4}/1.63 \times 10^{-4} = 1.65 \approx \sqrt{2}$. This can be understood from Eqs. (3.30) and (3.31) as follows. Assume the random error in $A_{1,i}$ and $A_{2,i}$ are independent of each other with the same standard deviation σ_A. According to Eq. (3.30), the random error in measured beta function is

$$\sigma_\beta = \sqrt{(\partial\beta/\partial A_{i,1})^2 + (\partial\beta/\partial A_{i,2})^2}\,\sigma_A = 2\sqrt{\mathcal{F}\beta}\,\sigma_A. \tag{3.35}$$

From Eq. (3.31), the random error in measured phase function is

$$\sigma_\psi = \sqrt{(\partial\psi/\partial A_{i,1})^2 + (\partial\psi/\partial A_{i,2})^2}\,\sigma_A = \sqrt{\frac{\mathcal{F}}{\beta}}\,\sigma_A. \tag{3.36}$$

Since $\Delta\psi$ is the difference between two measured phase functions,

$$\sigma_{\Delta\psi} = \sqrt{2}\sigma_\psi = \sqrt{\frac{2\mathcal{F}}{\beta}}\sigma_A = \frac{1}{\sqrt{2}}\frac{\sigma_\beta}{\beta}. \tag{3.37}$$

Figure 3.28 shows the estimated rms measurement errors for different random BPM calibration errors with σ_{cal} ranging from 0 to 5.0%. As expected from Eq. (3.31),

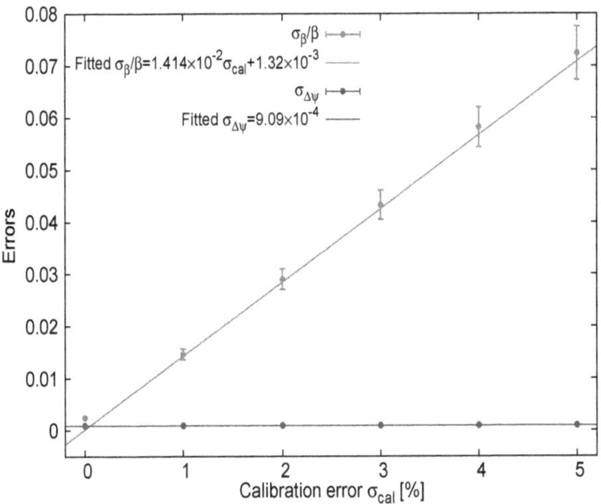

Figure 3.28: Estimation of rms measurement errors σ_β/β and $\sigma_{\Delta\psi}$ with various random BPM calibration errors σ_{cal}.

the measurement of the phase function is independent of BPM calibration errors. With a realistic BPM calibration error of 1.0% [37], the rms beta function measurement error $\sigma_\beta/\beta \approx 1.5\%$, such that the residual rms measurement error due to the averaging method is negligible.

Since its strong robustness for optics measurement had been verified by computer simulation, the SOBI ICA algorithm was applied to the TBT BPM data of AC dipole driven oscillation for a beam optics measurement and correction experiment in the

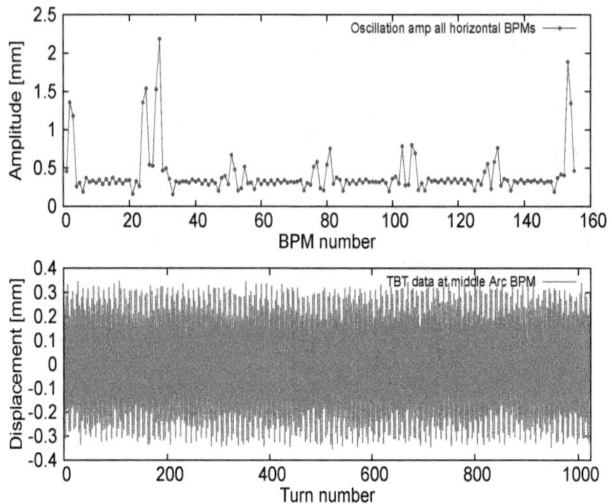

Figure 3.29: Amplitude of AC dipole driven betatron oscillation
at all available horizontal BPMs (top) and TBT data
at a BPM in the middle of the arc (bottom).

RHIC 2013 polarized proton operation. The experiment was taken at the storage stage with 255 GeV proton beams. Since the horizontal and vertical AC dipoles at RHIC are common to both rings, the bare betatron tunes of both rings are set to the same as $(\nu_x, \nu_z) = (28.6914, 29.6853)$ to facilitate simultaneous optics measurement for both rings. Horizontal and vertical AC dipoles were fired separately with $\Delta\nu = \pm 0.008$ to excite sustained driven betatron oscillation. TBT data for 1024 turns was then acquired at all available BPMs. The top plot of Fig. 3.29 shows the amplitudes of the AC dipole driven betatron oscillation at all available horzintal BPMs, in which the large amplitudes occurred at the BPMs close to the the triplet quadrupoles where the beta functions are large. The bottom plot of Fig. 3.29 shows an example of TBT data at a BPM in the middle of the arc with an amplitude about 0.3 mm.

The TBT BPM data were then analyzed by the SOBI ICA algorithm. The modes corresponding to driven betatron oscillaiton were extracted, as shown in Fig. 3.30. The temporal function reveals the tune distance is $\Delta\nu = 0.008$. The spatial functions were used to calculate the beta functions and beta-beat.

Figure 3.31 shows the measured beta-beat for both rings. In the horizontal direction, beat-beat was distributed smoothly along both ring. The horizontal peak beta-beat in both rings were approximately 15%. In the vertical direction, the peak beta-beat reached 30% in the Blue ring and 60% in the Yellow ring.

3.5 Summary

In this chapter, RHIC optics and instrumentation are first introduced. Selected optics measurement techniques are then briefly reviewed. Next, the principles of a time-correlation based ICA algorithm called SOBI are discussed. At last, simulation and experimental results of the SOBI ICA algorithm applied to RHIC for estimation of BPM noise performance and optics measurement from TBT BPM data of driven

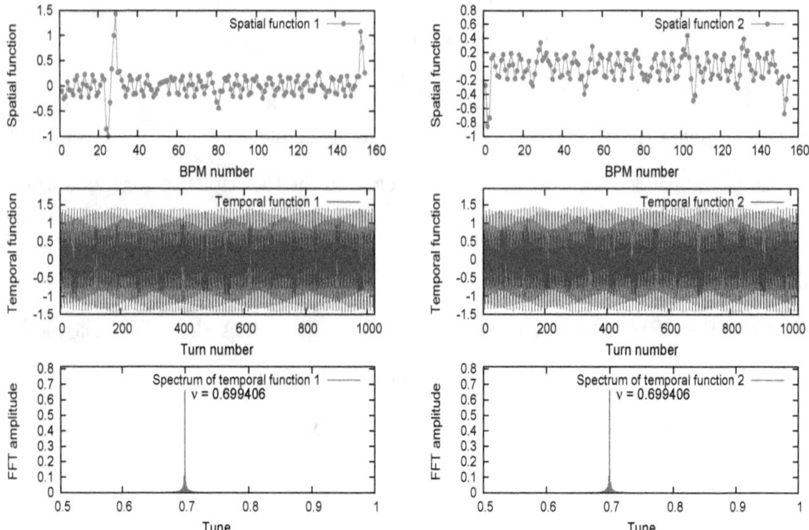

Figure 3.30: Spatial function (top), temporal function (middle) and spectra of temporal function for the first (left) and second (right) modes corresponding to driven betatron oscillation. The units for all vertical axes are arbitrary.

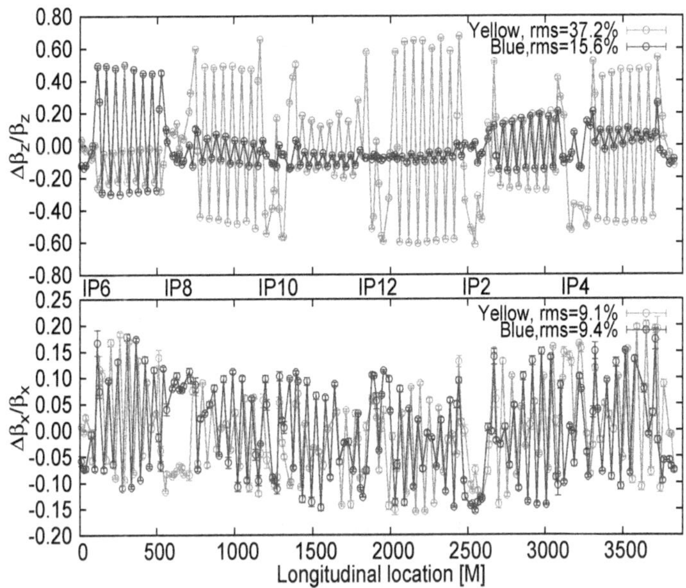

Figure 3.31: Measured horizontal (bottom) and vertical (top) beta-beat for both rings at RHIC.

betatron oscillation are presented.

General ICA methods are based on a quantity to measure statistical independence and an algorithm to optimize the quantity. There are many quantities proposed for measurement of independence, such as nonlinear functions [61], fourth cumulant [60] and time-lagged covariance matrices [62]. Many of the optimization algorithms are based on either orthogonal or non-orthogonal joint diagonalization of a set of matrices [70]. By adopting the time-lagged covariance matrices as a criterion for independence and a Jacobi-like orthogonal joint diagonalization algorithm for optimization, the SOBI ICA algorithm provides efficient and robust separation of narrow-band beam oscillation signals from TBT BPM data in a synchrotron. In Ref. [71], an ICA algorithm using non-orthogonal joint diagonalization on the Stiefel Manifold shows better robustness against noise than the SOBI algorithm. This algorithm is worthwhile for future study towards more contaminated data. Cumulant-based ICA algorithms, such as JADE [60], may also be useful for analysis of data that does not have explicit time correlation in an accelerator, such as shot-by-shot BPM data in a linear accelerator.

The high efficiency of the SOBI ICA algorithm makes it possible for online analysis of BPM noise based on parasitically acquired TBT BPM data of quiet beams. Perturbative collective beam oscillation signal can be extracted for inspection of harmful large beam collective oscillation, while the routinely estimated BPM noise provide useful logged information for BPM performance supervision and maintenance. Figure 3.32 shows the BPM noise for different bunch intensities estimated from parasitic TBT BPM data for the Yellow ring for the 2014 Au-Au operation. As expected, the overall BPM noise decreases when the bunch intensity increases.

The SOBI ICA results also contains source signals other than those for the betatron oscillation. Figure 3.33 shows an example of the synchrotron mode extracted from a TBT BPM data of AC dipole driven betatron oscillation taken during the RHIC 2013 polarized proton operation. The large excursion in the spatial function

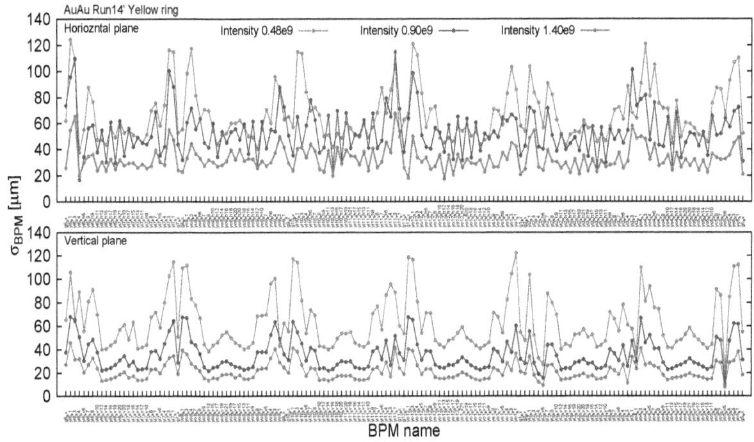

Figure 3.32: Estimated BPM noise for different bunch intensities for the Yellow ring from parasitically measured TBT BPM data for the 2014 Au-Au operation.

occurs at IP 4 where the RF cavities are located. Since only 1024 turns are acquired, the FFT resolution for tune is only $1.0/1024 = 9.766 \times 10^{-4}$ such that the synchrotron tune $\nu_s = 5.0 \times 10^{-4}$ for this polarized proton lattice can not be resolved. However, about one half of an oscillation within 1024 turns is observed from the temporal function, which shows a strong evidence for this mode to be corresponding to the synchrotron motion. The temporal function is related to fractional momentum deviation δ, while spatial function is related to the dispersion function D. Further investigations are needed to determine δ and D from the synchrotron mode.

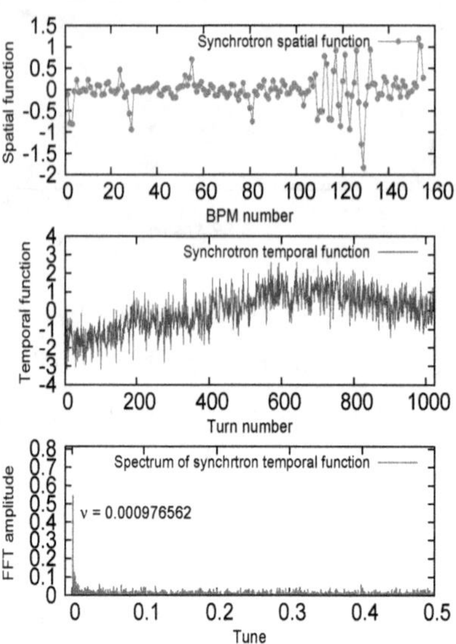

Figure 3.33: Spatial (top), temporal (middle) and FFT spectrum of temporal function (bottom) of the synchrotron mode from TBT BPM data of RHIC 2013 polarized proton operation.

Chapter 4

Optics correction at RHIC

The beta-beat may compromise beam stability, luminosity and polarization performance of RHIC. Optics measurement based on the SOBI ICA algorithm discussed in Chapter 3 shows large beta-beat at RHIC for the 2013 polarized proton operation. Therefore, it is highly desirable to develop optics correction schemes for routine operations in the control room for efficient minimization of beta-beat.

There are various methods for optics correction. The stopband compensation method relies on the fact that beta-beat is dominated by harmonics near twice the betatron tune [6]. The spectrum of the measured beta-beat at all BPMs are obtained by a Fourier analysis. A few families of trim quadrupoles are then excited to cancel the dominating harmonics [72]. Since the horizontal and vertical betatron tunes for the lattice of RHIC 2013 polarized proton operation are $(\nu_x, \nu_z) = (28.6914, 29.6853)$, horizontal and vertical beta-beat are dominated by the 57th to 60th harmonics, which cannot be accurately sampled by the only 160 BPMs available at RHIC. Therefore, the stopband compensation method is not efficient for beta-beat correction at RHIC.

The orbit response matrix (ORM) based lattice modeling technique such as the linear optics from closed orbit (LOCO) can also be used for optics correction. With the modeled lattice of the real machine, correction is calculated directly from the difference between the designed and the real lattice. Successful application of LOCO for optics correction are found in the spallation neutron source (SNS) at Oak Ridge National Lab [73], synchrotron light source SOLEIL [53], etc. LOCO was also proposed for RHIC [74], but a first trial of RHIC orbit response analysis with LOCO was not satisfactory [75] possibly because of the difficulty in fitting numerous parameters from limited number of BPMs for an accelerator as large as RHIC

The segment-by segment technique (SBST) was first invented at the Large Hadron Collider (LHC) for identification of large local gradient errors in the interaction regions (IRs) [76]. It was also successfully introduced to RHIC during the 2011 polarized proton operation and successfully reduced the peak beta-beat to a 20% level [77].

Due to the powering scheme at RHIC, only the triplet and trim quadrupoles in the IRs are independently powered. To fully utilize these available quadrupole correctors for global beta-beat correction, a scheme based on the beta-beat response matrix was systematically studied. To further reduce beta-beat in the arcs where there are no independently powered quadrupoles, a new scheme of using horizontal closed orbit bump at sextupoles in the arcs was explored. These correction schemes were successfully demonstrated during the 2013 RHIC polarized proton operation. Principles, simulation and experimental results of beta-beat response matrix global beta-beat correction method and the horizontal closed orbit at sextupoles for arc beta-beat correction scheme are presented in Section 4.1 and 4.2, respectively.

4.1 Beta-beat response matrix correction method

According to Eq. (2.48), beta-beat is a linear superposition of individual perturbation of integrated strength $\Delta K_1 L$ from all quadrupoles. Therefore, a beta-beat response matrix $\mathbf{R}$ can be defined as

$$\mathbf{B} = \mathbf{R}\mathbf{K},$$

(4.1)

where the observable vector $\mathbf{B}^T = (\frac{\Delta\vec{\beta}_x}{\beta_x}, \frac{\Delta\vec{\beta}_y}{\beta_z}, \Delta\nu_x, \Delta\nu_z)$ is composed of the beta-beat vectors $(\frac{\Delta\vec{\beta}_x}{\beta_x}, \frac{\Delta\vec{\beta}_y}{\beta_z})$ measured at the BPMs as well as the tune variations $(\Delta\nu_x, \Delta\nu_z)$, and the variable vector $\mathbf{K}^T = ([\Delta K_1 L]_1, [\Delta K_1 L]_2, \ldots, [\Delta K_1 L]_N)$ represents the change of integrated strength of N quadrupoles. In this discussion, N quadrupoles are used to correct M observables in $\mathbf{B}$. In the case with limited number of quadrupole correctors, i.e., $M > N$, Eq.(4.1) describes an overdetermined system in which beta-beat is minimized. To preserve beam stability and polarization, the tune variations must be limited to $\Delta\nu_{x,z} \leq 5 \times 10^{-4}$. In practice, uncalibrated BPMs should be excluded from the measurement, and the beta-beat measured from noisy BPMs should be assigned low weights. The tune variations must receive high weight in order to avoid large tune shifts which can cause excessive beam loss as well as polarization loss. Hence, different weighting factors are applied to Eq.(4.1) as:

$$\mathbf{W}\mathbf{B} = \mathbf{W}\mathbf{R}\mathbf{K},$$

(4.2)

where

$$\mathbf{W} = \begin{pmatrix} w_1 & 0 & \ldots & 0 \\ 0 & w_2 & \ldots & 0 \\ \vdots & \vdots & \ddots & \vdots \\ 0 & 0 & \ldots & w_M \end{pmatrix},$$

(4.3)

Each diagonal element w_i is the weighting factor for different BPMs and variations of tunes. The required correction strengths are computed by inverting the weighted response matrix $\mathbf{W}\mathbf{R}$ to solve Eq.(4.2). Singular value decomposition (SVD) can be applied to decompose $\mathbf{W}\mathbf{R}$ as

$$\mathbf{W}\mathbf{R} = \mathbf{U}\mathbf{\Gamma}\mathbf{V}^T,$$

(4.4)

where $\mathbf{U}$ is a real orthogonal $M \times M$ matrix, and $\mathbf{V}^T$ is a real orthogonal $N \times N$ matrix. $\boldsymbol{\Gamma}$ is a real diagonal $M \times N$ matrix with singular values $\Gamma_{11} = \gamma_1 \geq \Gamma_{22} = \gamma_2 \geq \cdots \geq 0$. To remove singularities, a tolerance level γ_c is chosen such that $\gamma_i \leq \gamma_c$ for $i > r$, where r is called the rank of $\mathbf{WR}$. Once the SVD of $\mathbf{WR}$ is obtained, the generalized inverse $(\mathbf{WR})^\dagger$ is given by

$$(\mathbf{WR})^\dagger = \mathbf{V}\boldsymbol{\Gamma}^{-1}\mathbf{U}^T, \tag{4.5}$$

where $\boldsymbol{\Gamma}^{-1}$ is a diagonal matrix with $\Gamma_{11}^{-1} = 1/\gamma_1, \cdots, \Gamma_{rr}^{-1} = 1/\gamma_r$ and 0 for all diagonal elements with $i > r$. The required correction strengths for global beta-beat correction $\mathbf{K}_{\mathrm{cor}}$ are computed as

$$\mathbf{K}_{\mathrm{cor}} = -(\mathbf{WR})^\dagger\mathbf{WB}. \tag{4.6}$$

The correction given in Eq.(4.6) is equivalent to a solution of an overdetermined weighted χ^2 minimization of the beta-beat in which the 2-norm of $\mathbf{K}_{\mathrm{cor}}$ is minimized.

During the RHIC 2013 polarized proton operation, the beta-beat response matrix based correction method was demonstrated in an experiment with 2 hours beam time. Table 4.1 summarizes the parameters for the beta-beat correction experiment in both rings. A total of 72 triplets and trim quadrupoles in the interaction regions with independent power supplies are used as beta-beat correctors, while two families of arc quadrupoles called "QF" and "QD" are also included as tune correctors. Therefore, there are 74 elements in the variables vector $\mathbf{K}_{\mathrm{cor}}$ in a sequence such that the first 72 elements are the IR quadrupole correctors starting from IP6 and arranged in a clockwise manner, and the 73rd and 74th elements are QF and QD, respectively.

With an assumption that the beta-beat originates from perturbations, the response matrix $\mathbf{R}$ was numerically computed from the ideal lattice by individually perturbing the integrated strength of each quadrupole corrector and recording the unit response of beta-beat and tune shifts. Figure 4.1 shows a surface plot of the response matrix, where the sequence of the variables are the same as that of $\mathbf{K}_{\mathrm{cor}}$ and

Table 4.1: Parameters for optics measurement and correction

Parameter	Blue	Yellow
Horizontal tune ν_x	28.6914	28.6914
Vertical tune ν_z	29.6853	29.6853
Number of horizontal BPM (used/total)	157/160	156/160
Number of vertical BPM (used/total)	158/160	158/160
Number of beta-beat correctors	72	72
Number of tune corrector families	2	2

the order of the observables are identical to that of **B**. As expected, large response

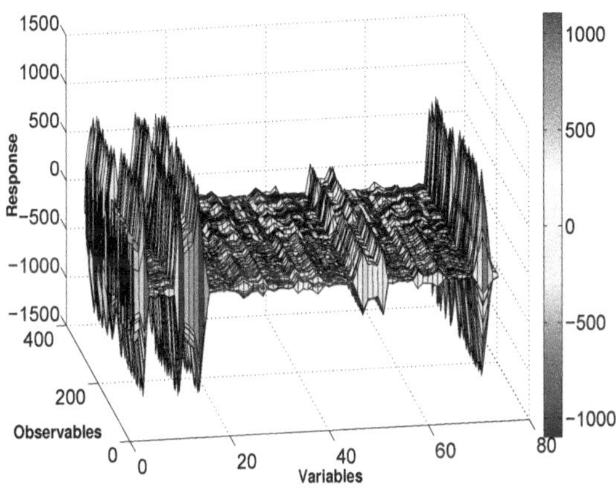

Figure 4.1: Surface plot of the beta-beat response matrix versus

occurs at correctors at locations with large beta functions. The phase advance between each quadrupole in the QF or QD family is close to $\pi/2$ such that each family has small contribution to beta-beat but large influence to tune variations, as shown in Fig. 4.2. Therefore, QF and QD are mainly used to limit tune variations.

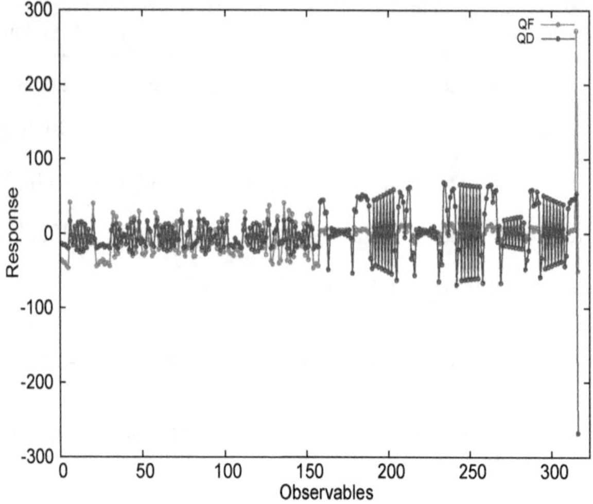

Figure 4.2: Beta-beat response of QF and QD.

The correction $\mathbf{K}_{\mathrm{cor}}$ was computed by scanning the weighting factors and rank of response matrix $\mathbf{R}$ to find a best solution that reduces the global beta-beat as much as possible with minimum tune variations. For example, Figure 4.3 shows the dependence of tune variations and residual beta-beat on the weighting factor and rank of the response matrix, from which the solution which obeys the limit $\Delta\nu_{x,z} < 2\times10^{-3}$ and minimizes the residual beta-beat can be found.

In event of large beta-beat, the response matrix calculated from the ideal lattice will apparently deviate from the true response in the machine. However, with successive application of correction based on Eq. (4.6), the response matrix from ideal lattice will be closer to the true response in the machine, and beta-beat will be gradually reduced.

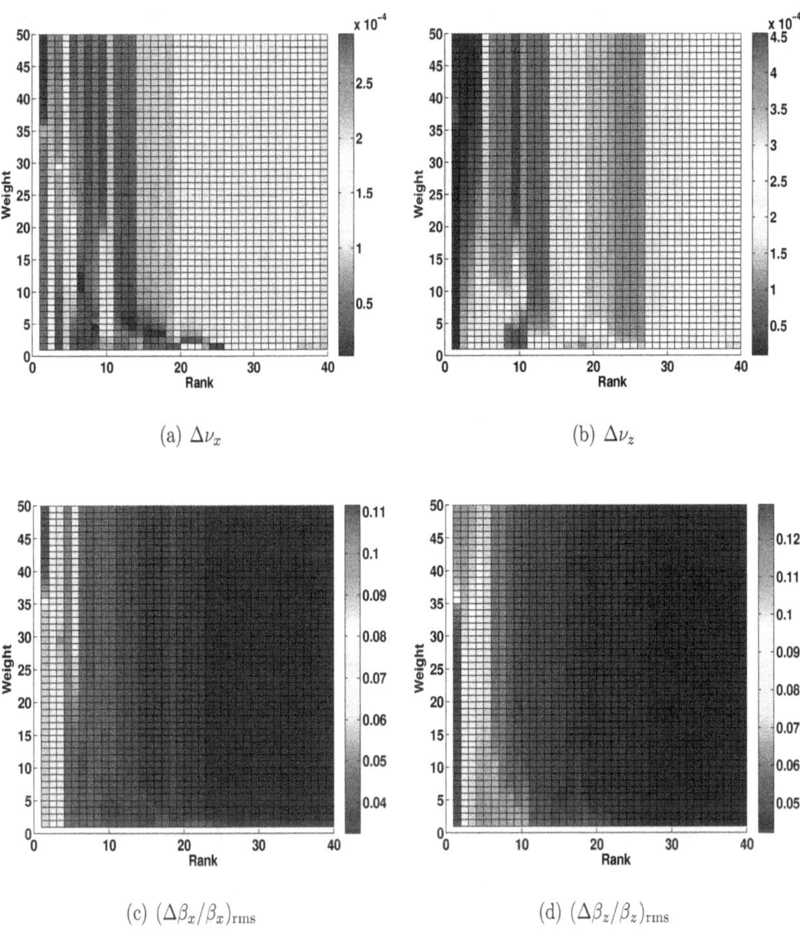

(a) $\Delta\nu_x$

(b) $\Delta\nu_z$

(c) $(\Delta\beta_x/\beta_x)_{\mathrm{rms}}$

(d) $(\Delta\beta_z/\beta_z)_{\mathrm{rms}}$

Figure 4.3: Dependence of tune variation and rms residual beta-beat in the horizontal (left) and vertical (right) directions on the weighting factor and rank of the response matrix.

Figure 4.4 shows the computed relative correction strengths $\Delta(K_1L)/K_1L$ for the Blue ring. All of the relative corrections are within 1.0%. The relative changes in many trim quadrupoles are large because trim quadrupoles are normally set at a low field. Figure 4.5 shows a simulation of the evolution of tunes and rms beta-beat

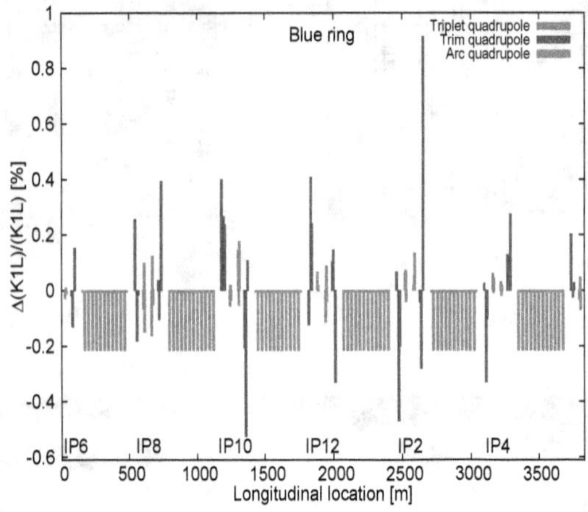

Figure 4.4: Relative changes of quadrupole integrated strength as a function of quadrupole locations in the Blue ring.

in the Blue ring along a ramp-up process of the computed correction. There are no tune variations after 100% correction strength. During the ramp-up process, the excursions of tune changes are within 5×10^{-4}. The rms beta-beat in both directions is reduced smoothly during the correction ramp-up process. The small increase of horizontal rms beta-beat at the end of the correction ramp-up process is because of decreasing vertical beta-beat and tune compensation. The measured rms beta-beat in Fig. 4.5 shows a good agreement with the predicted values.

Figure 4.6 shows the measured beta-beat with and without correction for the Blue ring. The horizontal peak beta-beat was successfully reduced from 15% to 8%. In

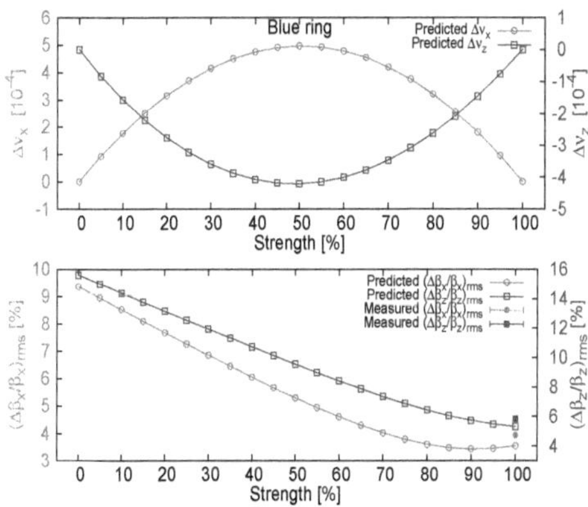

Figure 4.5: Simulated tune variations (top) and residual rms beta-beat (bottom) versus correction strength for the Blue ring. The measured rms beta-beat with error bars at 100% correction strength is also shown.

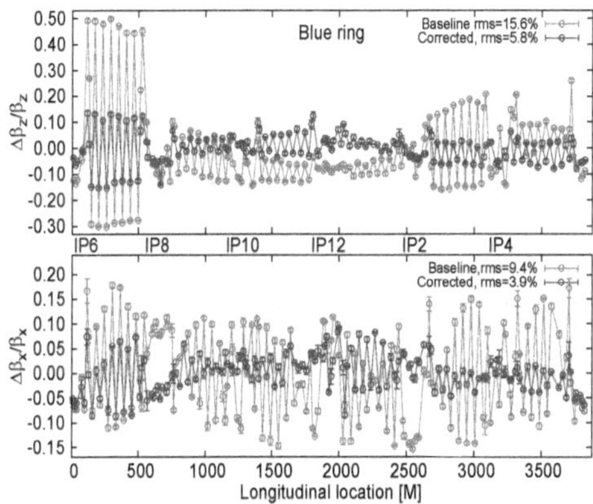

Figure 4.6: Baseline and corrected horizontal (bottom) and vertical (top) beta-beat with error bars for the Blue ring.

the vertical direction, significant suppression of beta-beat was also achieved in the arc between IP6 and IP8 as well as the one between IP2 and IP4. The vertical peak beta-beat was reduced from 40% to 14%.

The correction in the Yellow ring which has excessively large beta-beat in the vertical direction is more involved. The relative correction strengths for the first trial of corrections are shown in Fig. 4.7 by the hollow bars. All relative correction strengths are within 0.6%. The large relative correction strengths in the arc quadrupoles were due to compensation of the large tune shifts caused by the triplet and trim quadrupoles employed to minimize the beta-beat. Figure 4.8 shows the

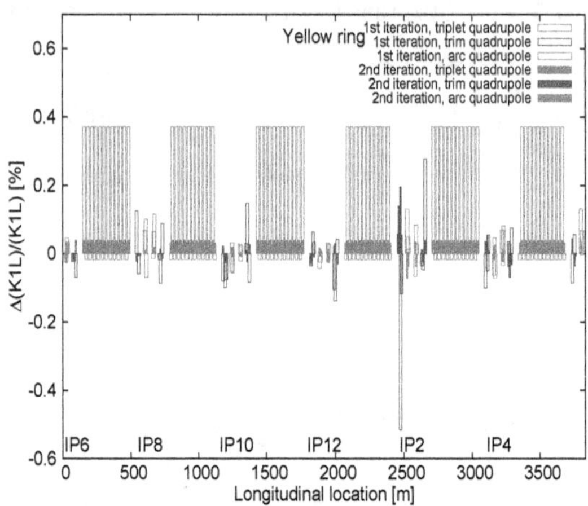

Figure 4.7: Relative changes of quadrupole integrated strength in the Yellow ring for the first iteration (hollow bars) and second iteration (solid bars).

simulated evolution of tunes and rms beta-beat along the ramp-up process of the first correction as well as the measured rms beta-beat at 100% correction strength. The excursions of tune variations are within 2×10^{-3}. The rms beta-beat is reduced smoothly as correction strength increases. At full correction strength, the 5% mea-

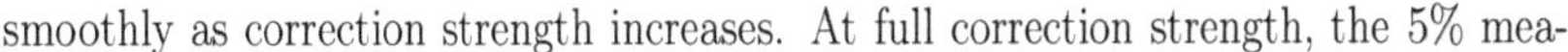

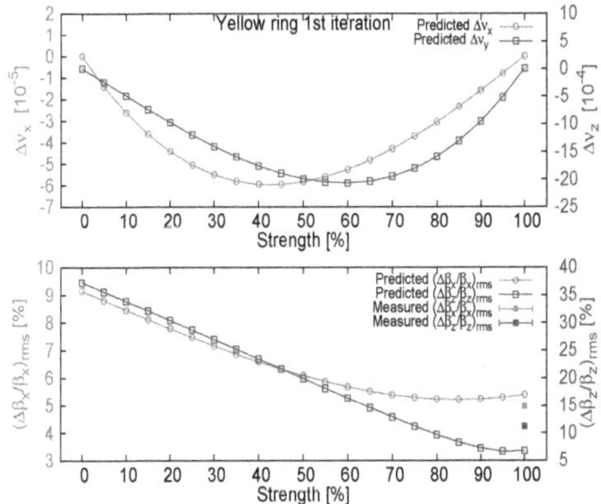

Figure 4.8: Simulated tune variations (top) and residual rms beta-beat (bottom) versus the strength of the first iteration of correction for the Yellow ring. The measured rms beta-beat with error bars at 100% correction strength is also shown.

sured horizontal rms beta-beat is lower than the predicted 5.4% value. The top plots in Figs. 4.9 and 4.10 show the measured beta-beat of the Yellow ring with and without the correction. The horizontal peak beta-beat was reduced to 12%. However, in the vertical direction there was still a peak beta-beat as large as 20%, and the 11% measured vertical rms beta-beat is about 2 times of the prediction shown in Fig. 4.8. This is due to the initial large beta-beat in the vertical direction. Hence, a second iteration was exercised.

The results of the second iteration of correction along with the first iteration are shown in the two bottom plots in Fig. 4.9 and 4.10. After the second iteration, significant vertical beta-beat reduction was achieved, and the peak beta-beat was

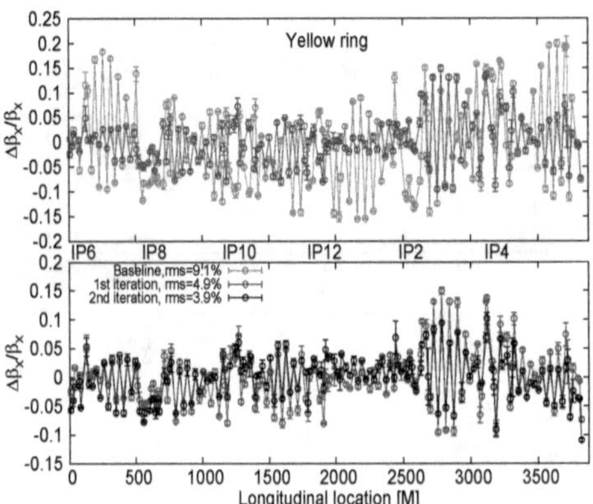

Figure 4.9: Baseline and corrected horizontal beta-beat with error bars for the Yellow ring.

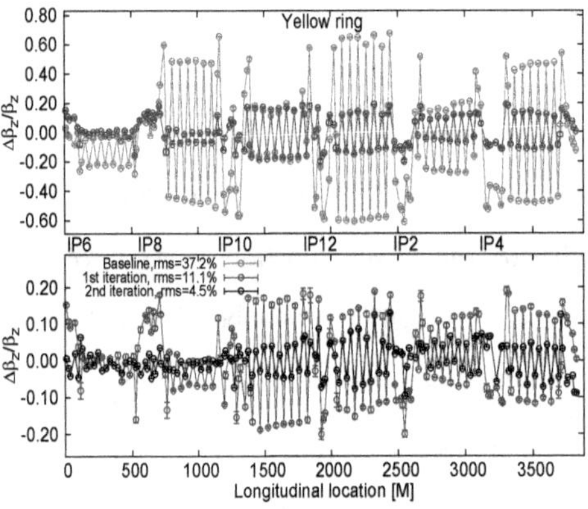

Figure 4.10: Baseline and corrected vertical beta-beat with error bars for the Yellow ring.

successfully reduced to approximately 10% for both planes. The computed relative correction strengths are shown in Fig. 4.7 by the solid bars. The relative correction strengths are smaller than those for the first iteration. Similar to Fig. 4.8, the evolution of tunes and rms beta-beat for the second iteration was also computed and shown in Fig. 4.11. The excursions of tune variations are within 4×10^{-4} and the changes of rms beta-beat are smooth. At full correction strength, the 4% measured horizontal rms beta-beat matches the predicted value of 3.8%, and the deviation between measured and predicted vertical rms beta-beat is 0.6% after the second iteration.

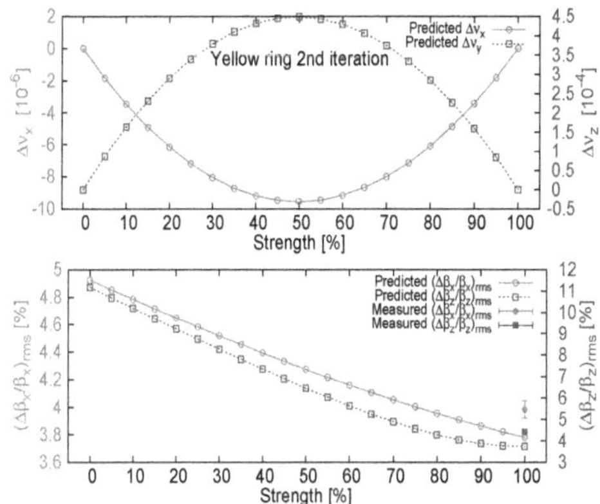

Figure 4.11: Simulated tune variations (top) and residual rms beta-beat (bottom) versus the strength of the second iteration of correction for the Yellow ring.The measured rms beta-beat with error bars at 100% correction strength is also shown.

It is expected for a successful optics correction scheme that the minimization of beta-beat should also minimize the relative phase-beat $\Delta\psi_{ij}$. Thanks to the high quality BPMs at RHIC and on-going efforts in further improving BPM performance over a decade of RHIC operation, our optics correction based on beta-beat response matrix also results in significant reduction in relative phase-beat between consecutive BPMs, especially in the vertical plane, which is shown in Fig. 4.12. This also validates the proposed optics correction method.

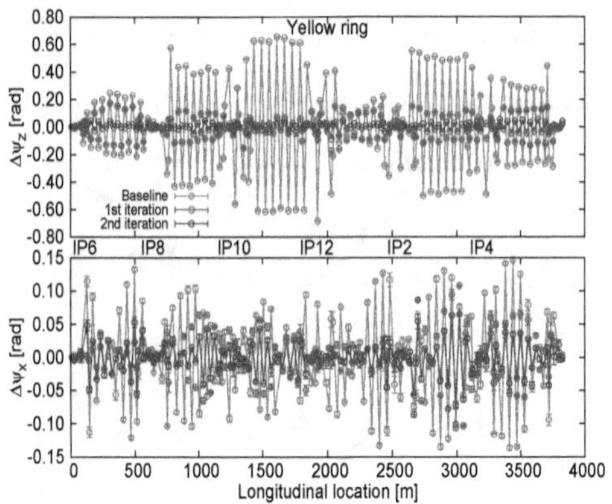

Figure 4.12: Baseline and corrected horizontal (bottom) and vertical (top) relative phase-beat with error bars for the Yellow ring.

Due to the difference between the real machine and the ideal lattice, a variation of the tunes on the order of 10^{-3} was observed after the corrections. This can be corrected afterward. In conclusion, the proposed optics correction method based on beta-beat response matrix has successfully reduced both horizontal and vertical peak beta-beats to 10% in both rings. Good agreement was also found between measurements and predictions. There were no possibilities to apply additional iterations of correction in this experiment due to the limited beam time. For further correction of arc beta-beat, additional techniques are needed.

4.2 Arc beta-beat correction using closed orbit bump and sextupole

In high energy accelerators, very often the number of available arc quadrupoles with independent power supplies is limited. In RHIC, all arc quadrupoles are grouped into two families called QF and QD, and each family is powered by a common power supply. Therefore, gradient errors in the arcs cannot be effectively compensated with global correction using only quadrupoles in the IRs. However, according to Eq. (2.13), a non-zero horizontal closed orbit x_{co} at a sextupole results in magnetic field

$$B_x = B_2 x_{\mathrm{co}} z, \quad B_z = B_2 x_{\mathrm{co}} x. \tag{4.7}$$

This is exactly the same as normal quadrupole field with an equivalent quadrupole field gradient $B_2 x_{\mathrm{co}}$. Therefore, this feed-down quadruple field can be used to correct arc beta-beat [78]. For RHIC, horizontal closed orbit bumps at the sextupoles in the arcs can be excited by dipole correctors to create the feed-down quadrupole field to provide additional beta-beat correctors in the arcs. Once the response matrix of beta-beat to the amplitude of the horizontal closed orbit bump is obtained from the ideal model, an SVD inversion method discussed in Section 4.1 can be applied to compute the required pattern of closed orbit bumps. Demonstration of this technique is of great interest for a high energy accelerator like RHIC, as well as LHC, where precise control of linear optics is required to facilitate optics manipulations. Such an example can be the Achromatic Telescopic Squeezing (ATS) [79] for further increasing of luminosity.

A proof-of-principle experiment was carried out in the Yellow ring at beam energy of 255 GeV during the RHIC polarized proton run in 2013. SBST was first applied to correct large local beta-beat. The beta-beat response matrix based method using quadrupoles in interaction regions and arcs was applied to reduce peak beta-beat in both transverse directions to approximately 10% to facilitate arc beta-beat correction. The computed horizontal closed orbit bumps were then applied by the closed orbit feedback system [45]. Figure 4.13 shows the computed horizontal closed orbit required for arc beta-beat correction. The measured horizontal closed orbit in Fig. 4.13 shows a

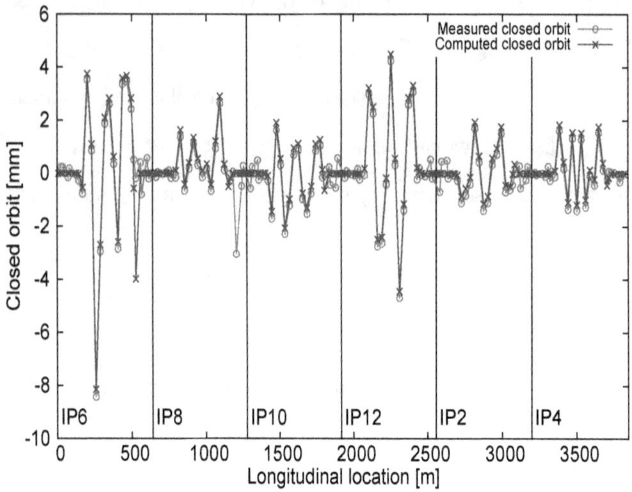

Figure 4.13: Measured and computed horizontal closed orbit for correction of arc beta-beat in the Yellow ring.

good agreement with prediction in all arcs. At around $s = 500$ m, the measured closed orbit missed a computed orbit bump. This is because in the computed closed orbit three horizontal orbit correctors were used to match the closed orbit at the beginning of the interaction region at IP8. But, in the experiment the orbit feedback system used more than three horizontal orbit correctors and resulted in smaller bumps at the same location. However, these orbit bumps do not affect the beta-beat correction because there are no sextupoles in this location. The large closed orbit bump measured at around $s = 1200$ m near IP10 is intrinsic to RHIC for beam dump. This bump is not considered in the computed closed orbit. However, it does not affect the correction results since there are no sextupoles in this region as well. No beam loss was observed with this closed orbit pattern in which a maximum excursion about 8.5 mm was observed in between IP6 and IP8. The correction results are shown in Fig. 4.14. In

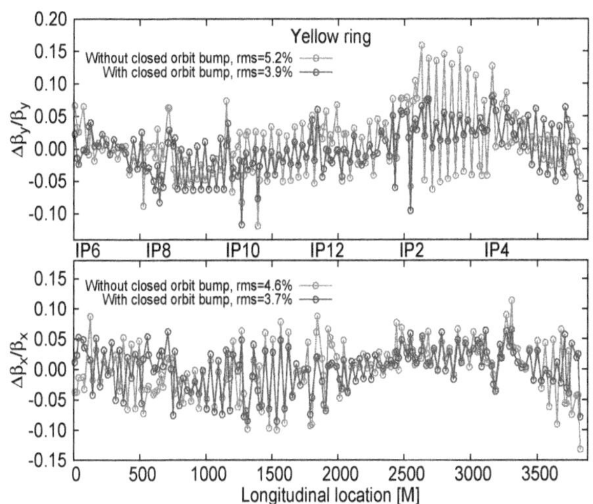

Figure 4.14: Horizontal (bottom) and vertical (top) residual beta-beat with and without the horizontal closed orbit bump displayed in Fig. 4.13 for the Yellow ring.

the horizontal plane, beta-beat reduction was seen clearly in the arc between IP10 and IP12 as well as the arc between IP4 and IP6, while beta-beat in the other arcs remains about the same. Peak beta-beat was successfully reduced to approximately 7%. In the vertical plane, significant beta-beat reduction was observed in the four arcs between IP10 and IP4. Especially in the arc between IP2 and IP4, peak beta-beat was remarkably reduced from 10% to 4%. Overall, the vertical peak beta-beat was reduced from 10% to 7%.

Due to limited beam time, only one set of measurements was taken such that no error bars are available for this beta-beat. However, experiences of previous measurements show that a typical average value of error bar is about 0.5%. There were not any opportunities to further explore beta-beat correction using larger horizontal closed orbit bumps within limited beam time. Nonetheless, the successful demonstration of using horizontal closed orbit bump at sextupoles to correct arc beta-beat at RHIC polarized proton store energy shows this technique is feasible for high energy accelerator operations.

Figure. 4.14 also shows a modulation of the measured beta-beat such that the offset of beta-beat varies similarly as a sinusoidal wave with a period equal to the circumference of the ring. The amplitude of the variation is about 2%. This modulation effect may have resulted from distributed coupling errors, such as triplet quadrupole roll errors and skew quadrupole errors. Further investigations are needed to identify the sources of this modulation effect.

4.3   Summary

In this chapter, theory and experimental results for optics correction schemes developed for RHIC during the 2013 polarized proton operation are presented. The beta-beat response matrix based global correction method and arc beta-beat correction using horizontal closed orbit at sextupoles were systematically studied and successfully demonstrated in beam experiments. With the accurate optics measurement technique introduced in Chapter 3 and the efficient optics correction schemes, a record-low peak beta-beat of 7% was experimentally achieved at RHIC. This result is also taking the leading level in the achieved peak beta-beat among various high energy colliders, as shown in Table 4.2.

The beta-beat response matrix based method can be easily expanded for simul-

Table 4.2: Achieved peak beta-beat of various lepton (top half) and hadron (bottom half) colliders [2]

Collider	Circumference [km]	Peak $\Delta\beta/\beta$ [%]
PEP II	2.2	30
LEP	27	20
KEKB	3	20
CESR	0.8	7
HERA-p	6.3	20
Tevatron	6.3	20
LHC	27	7 (2012) [80]
RHIC	3.8	20 (2012) / 7 (2013)

taneous correction of more optics parameters such as the dispersion function. For example, in Fig. 4.15, the red curves shows simulated perturbations in beta and dispersion functions, while the blue curves shows the results reproduced by the response matrix based optics correction method. The reproduced results successfully capture the features of the perturbations. The correction is applied by inverting the sign of the correction strength and the perturbations are expected to be minimized. This example sets up a foundation for more comprehensive optics correction studies in the future.

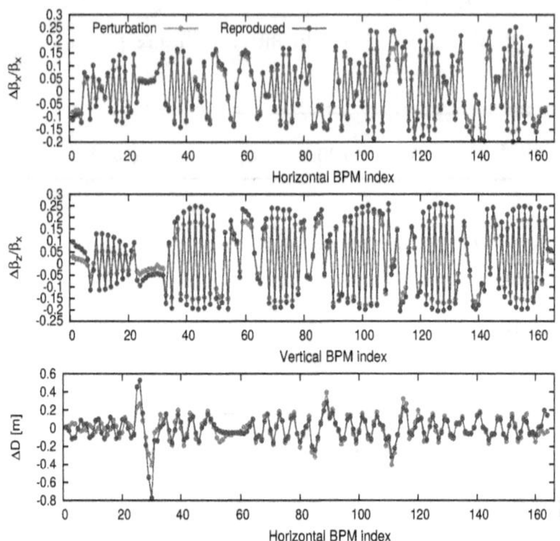

Figure 4.15: Simulated perturbations in beta and dispersion functions and the reproduced results.

Chapter 5

Software packages

One of the most important missions for the topic of optics measurement of correction at RHIC is to develop computer software packages which are both consistent with the RHIC control system and easily adaptable to different accelerators. Since the object oriented programming language C++ provides high execution efficiency, rigorous data structure and flexible platform adaptability, it is used to develop a series of shared libraries to facilitate computer simulations and data analysis. To minimize the user input for optics measurement, a graphical user interface (GUI) was developed using ROOT [81] which is built on the C/C++ framework. Discussions of the C++ libraries are presented in Sec. 5.1. The GUI for optics measurement is introduced in Sec. 5.2.

5.1   C++ shared libraries

The following C++ shared libraries are powered by Armadillo [82] for linear algebra calculations. The source codes are available online at `http://pages.iu.edu/~xiaoshen/`.

Blind source separation (BSS) library

As discussed in Sec. 3.4, independent component analysis (ICA) for BSS problem is composed of a cost function to measure independence and an algorithm to optimize the cost function. The BSS library provides two cost functions, time-lagged covariance matrices and fourth cumulant. It implements a Jacobi-like algorithm which is an orthogonal joint diagonalization algorithm, and non-orthogonal joint diagonalization algorithms such as the least-squares diagonalization (LSDIAG) [83], fast Frobenius diagonalization (FFDIAG) [84], and joint diagonalization combining Givens and hyperbolic rotations [85].

The optics measurement method proposed in Section 3.4 was applied to RHIC through the BSS libraries. The other cost functions and joint diagonalization algorithms in the BSS libraries supply possibilities to explore their potential usage in accelerator data analysis. At the same time, the BSS library provides a generic data structure such that more cost functions and joint diagonalization algorithms can be easily integrated.

SimTrack library

SimTrack [65] is a C++ particle tracking code for beam dynamics study. It adopts the 4-th order symplectic integrator [86] to propagate particle phase space coordinates in a beam line. It is also integrated with the first order truncated power series algebra (TPSA) [87] to calculate the linear one-turn map from particle tracking such that linear optics parameters can be derived from the linear one-turn map [88]. SimTrack also provides the capability of weak-strong beam-beam effect simulation [89].

SimTrack was originally written as a C++ header file. To improve the code efficiency, it was rewritten into a C++ shared library with separated header and implementation files. A bug in adding nonlinear magnetic field components in a sector dipole magnet was also fixed [90]. The SimTrack library is used for particle tracking and optics calculation for simulations in this thesis. As a quick benchmarking, Fig. 5.1 shows the relative errors of beta functions calculated by SimTrack to those by MADX. This tiny discrepancy may arise from different tracking models used by SimTrack and MADX.

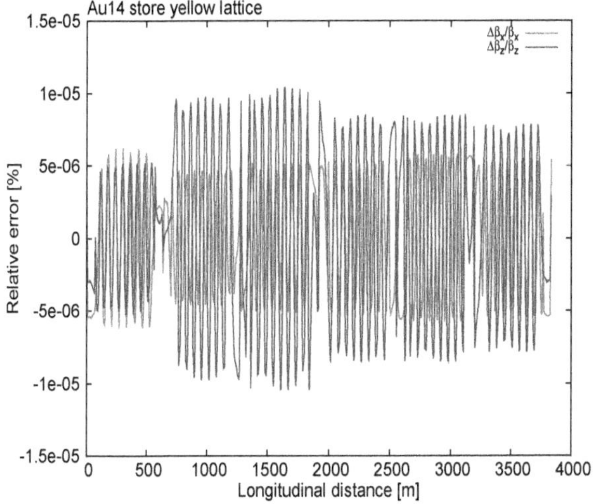

Figure 5.1: Relative errors of beta functions calculated by Sim-
Track to those by MADX.

The data architecture of the SimTrack libraries makes it simple to include additional accelerator elements such as undulator, and physical process of beam dynamics such as synchrotron radiation and beam cooling.

Numerical Analysis of Fundamental frequency (NAFF) library

The NAFF algorithm reviewed in Sec. 3.3 was implemented into a `C++` NAFF library for high resolution Fourier analysis. The NAFF library adopts a Hanning window to pre-process the raw data. Starting from the standard fast Fourier transform (FFT) as a first guess, NAFF uses an inverse quadratic interpolation method to find the best estimation for a selected number of dominating frequency components. Figure 3.14 shows a comparison of relative errors in amplitude A and frequency f

of a sinusoidal signal determined by the NAFF library and a standard FFT routine versus different number of signal length N, where NAFF shows an accuracy which is proportional to $1/N^4$.

In the optics measurement study for RHIC, NAFF is used to determine characteristic frequencies of source signals separated by ICA. NAFF is also useful in normal form analysis to measure the resonance driving terms (RDTs) for nonlinear beam dynamics study [91].

Geodesic accelerated Levenberg-Marquardt fitting library

The Levenberg-Marquardt method [92] is a nonlinear fitting algorithm which is widely used for various optimization problems. It has been applied in orbit response matrix (ORM) based accelerator lattice modeling techniques such as the linear optics from closed orbit (LOCO) [93]. Recently, an improved version called geodesic accelerated Levenberg-Marquardt algorithm was proposed [94] and showed better performance than the conventional method. This new algorithm was implemented into a geodesic accelerated Levenberg-Marquardt fitting C++ shared library. It is potentially useful for lattice modeling based on measured optics parameters such as the beta functions. For demonstration, a simulation was conducted in which the quadrupole strengths of a lattice composed of 12 FODO cells are perturbed and the geodesics accelerated Levenberg-Marquardt fitting library was used to fit an unperturbed lattice to perturbed one by varying the quadrupole strengths. Figure 5.2 shows the perturbed quadrupole strengths and fitted results. The geodesics accelerated Levenberg-Marquardt fitting library successfully captures all perturbations and accurately models the lattice.

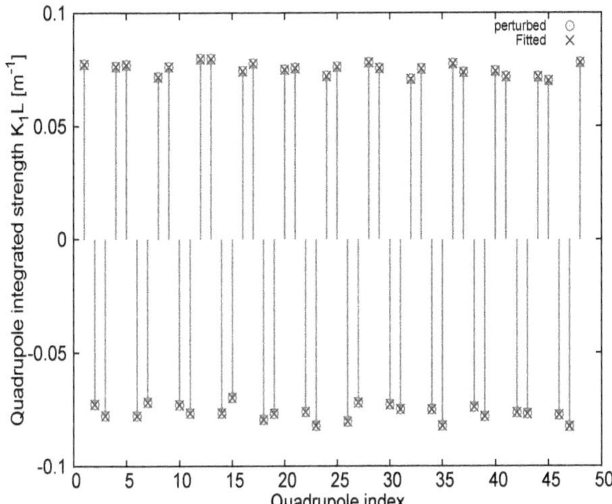

Figure 5.2: Perturbations in quadrupole integrated strengths and the fitting results by the geodesic accelerated Levenberg-Marquardt fitting library.

5.2   Graphical user interface

A GUI called **TbtAnalyzer** was developed using ROOT [81] to facilitate the TBT
BPM data analysis. There are two panels in **TbtAnalyzer**. The first panel is for raw
data pre-processing as shown in Fig. 5.3.

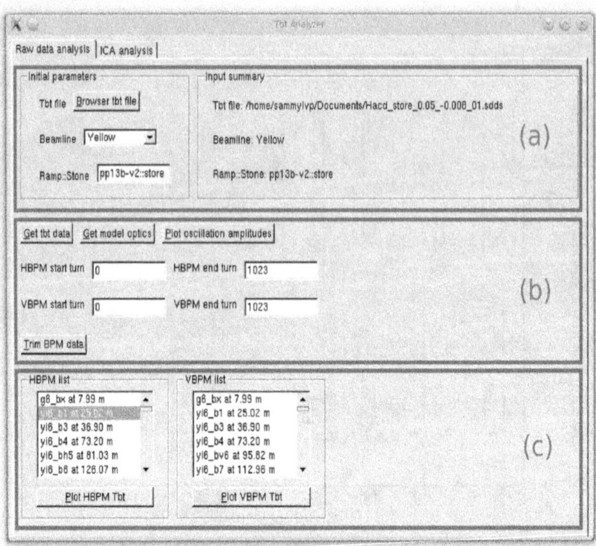

Figure 5.3: "Raw data analysis" tab of the GUI.

- The highlighted area (a) is for input of initial parameters including the raw TBT
  BPM file name, the beamline, and the code for the model optics parameters.

- The buttons in the highlighted area (b) execute commands appearing on them.
  The raw TBT BPM data whose file name is specified in area (a) is converted
  from a binary file with the self describing data sets (SDDS) format [95] into
  a human-readable ASCII format. At the same time, data with bad status are

excluded from the analysis. The model optics parameters are obtained through a background program from an online optics calculation engine OpticCal [96]. The beam oscillation can then be simply plotted, as shown by the top plot of Fig. 3.29. The BPM data can be trimmed by specifying the starting and ending turns.

- The highlighted area (c) shows lists of TBT BPM data sorted by their longitudinal distances to be plotted. Fig. 3.12 shows an example of the plotted TBT BPM data.

The second panel is for ICA analysis on the pre-processed data, as shown in Fig. 5.4

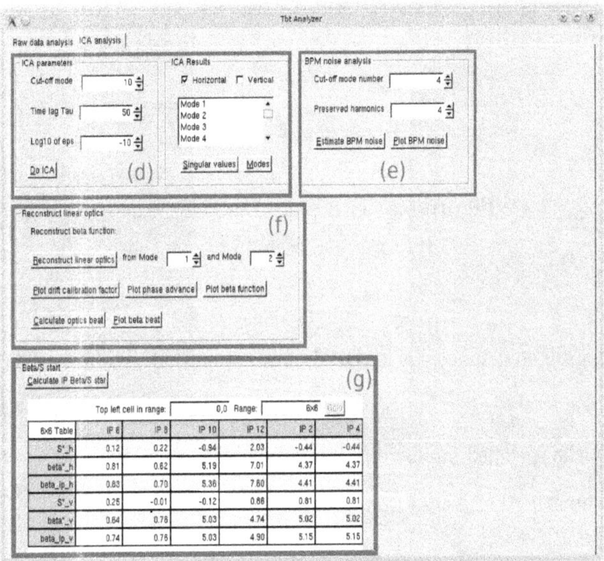

Figure 5.4: "ICA analysis" tab of the GUI.

- The highlighted area (d) is for input of SOBI ICA parameters including the cut-off mode number n_c, time lag parameter τ, and convergence accuracy for the Jacobi-like joint diagonalization algorithm. The singular values and modes composed of the spatial and temporal functions can also be plotted, as shown in Figs. 5.5 and 5.6, respectively.

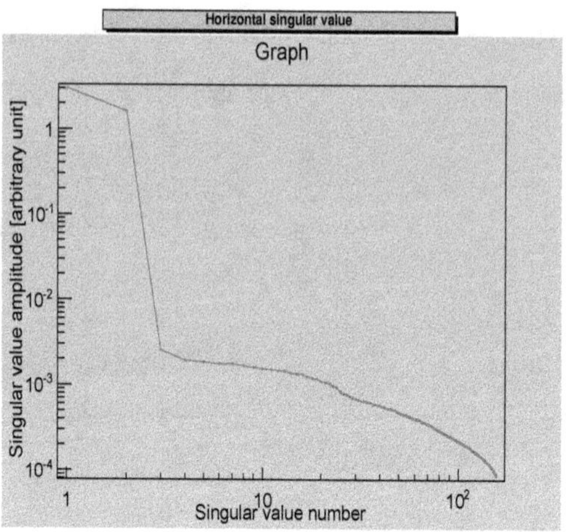

Figure 5.5: GUI plot of singular values.

- The highlighted area (e) is for BPM noise estimation. The cut-off mode number for noise analysis can be specified, and the NAFF algorithm is used to further purify the specified number of source signals by extracting the selected number of dominating harmonics from them.

- Once the betatron modes are identified from the highlighted area (e), beta functions and phase advances are reconstructed in the highlighted area (f) by specifying these modes. Corresponding plots can be simply drawn. The β^*, s^*, and β_{IP}, the beta functions at the IPs, are calculated using the measured beta

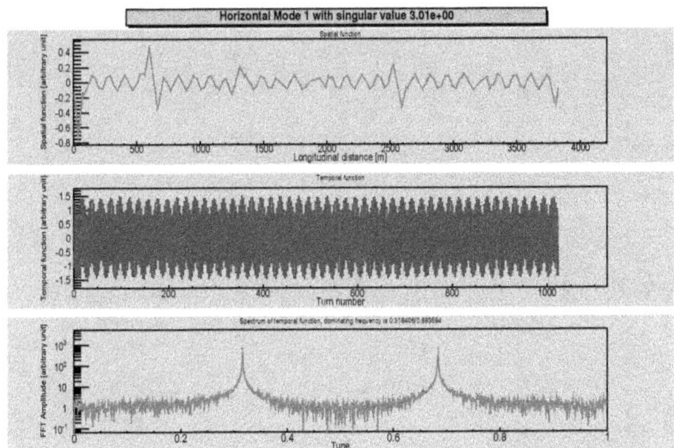

Figure 5.6: GUI plot of one mode.

functions from BPM close to the IPs and printed in the highlighted area (g).

5.3 Summary

In this chapter, software packages including the C++ shared libraries and a GUI developed for the mission of optics measurement and correction at RHIC are introduced. They are adapted to the RHIC data base and control system to guarantee the optics measurement and correction experiments was successfully finished within a beam time of only a few hours with satisfactory results.

The object oriented programming feature in the C++ libraries and GUI are simple to implemente and easily extended. For example, a TPSA of arbitrary order may be integrated into the SimTrack library to facilitate the normal form technique for nonlinear beam dynamics study [97]. Although the optics correction algorithms were written in separate codes, it can also be included in the GUI. These features set up a favorable foundation for future studies.

Chapter 6

Conclusion

In this thesis, optics measurement and correction techniques developed for the Relativistic Heavy Ion Collider (RHIC) are studied. The time-correction based independent component analysis (ICA) algorithm called second order blind identification (SOBI) was introduced to RHIC for beam position monitor (BPM) noise estimation and optics measurement. Efficient optics correction schemes were systematically studied and experimentally demonstrated in beam experiments. Software packages were developed for simulation and implementation of the proposed optics measurement and correction.

The turn-by-turn (TBT) BPM data of coherent beam motion excited by either a pulse kicker or an AC dipole can be considered as a linear mixture of source signals corresponding to physical beam motions with characteristic frequencies and BPM electronic noise. Extraction of the physical source signals provides useful information of beam dynamics. The SOBI ICA algorithm takes advantage of the time-correlation of the physical source signals to separate them by jointly diagonalizing the time-lagged covariance matrices of the whitened data. This algorithm was applied to analyze TBT BPM data at RHIC for a first systematic estimation of RHIC BPM noise performance.

The noise level was estimated to be lower than 65 μm for the 2013 polarized proton operation. The noise distribution over all BPMs showed a good agreement with the configuration of RHIC BPMs. An averaging method based on the SOBI ICA algorithm was developed for accurate linear optics measurement using TBT BPM data of AC dipole driven betatron oscillation. The averaging method was applied in a beam experiment during the 2013 polarized proton operation. The measured horizontal and vertical peak in the Blue ring were 15% and 30%, respectively, while those in the Yellow ring were 15% and 60%, correspondingly.

Efficient correction schemes are greatly desired to minimize the large beta-beat in both rings. In spite of the challenge that there were only a limited number of quadrupoles in the insertion region (IR) available as beta-beat correctors, a beta-beat response matrix correction method was developed for global beta-beat correction. This method was successfully demonstrated in a beam experiment and reduced the peak beta-beat to be 10% in both directions for both rings. Another scheme of using horizontal closed orbit bump at sextupoles was also proposed to make use of the feed-down normal quadrupole fields as additional beta-beat correctors in the arcs where there are no independently powered quadrupoles. A proof of principle experiment of this scheme applied in the Yellow ring achieved a record-low peak beta-beat of 7% in both directions.

To facilitate simulation study and experimental implementation of the proposed optics measurement and correction techniques, C++ shared libraries compatible with the RHIC database and control system were developed. A graphical user interface (GUI) was also developed to ease the TBT BPM data analysis. The object oriented programming feature makes it simple to extend the capabilities of these software packages.

Although this thesis was devoted to the study of RHIC, the optics measurement and correction techniques are applicable to general purpose beam diagnostics for circular accelerators. The software packages can also be simply adapted in the UNIX/Linux based operating system.